The Distribution of Income and Wealth

New Economic Windows

More information about this series at http://www.springer.com/series/6901

Fabio Clementi · Mauro Gallegati

The Distribution of Income and Wealth

Parametric Modeling with the κ-Generalized Family

Springer

Fabio Clementi
University of Macerata
Macerata
Italy

Mauro Gallegati
Polytechnic University of Marche
Ancona
Italy

ISSN 2039-411X ISSN 2039-4128 (electronic)
New Economic Windows
ISBN 978-3-319-27408-9 ISBN 978-3-319-27410-2 (eBook)
DOI 10.1007/978-3-319-27410-2

Library of Congress Control Number: 2015957801

Printed on acid-free paper

This Springer imprint is published by SpringerNature
The registered company is Springer International Publishing AG Switzerland

To Raffaella and Michela, who certainly deserve a better book than this one

Preface and Acknowledgments

In recent years, after a long period of decline, inequality in the distribution of income and wealth has increased. In the major OECD countries the Gini index went from 0.30 to 0.35; 0.1 % of the super-rich in the US in 2013 have 22 % of the country's wealth (note that from 1946 to 1988 the share was less than 10 %). To have similar levels one must go back to the years of the Great Depression.

The interest in income and wealth distribution is exemplified by the recent works of Stiglitz (2012, 2015a), Piketty (2014), and Atkinson (2015). It seems to have two separate causes: slow growth and its unsustainability. The first element is almost trivial: when the cake is growing a lot, one can legitimately worry less about its distribution. Conversely, now that millions of Europeans are at risk of poverty the problem of distribution arises strongly. And even more so when one considers global development, since it is incompatible with the current mode of production and consumption. Redistribution becomes a necessity.

The economic crisis has translated, also, into a crisis of economic theory. In particular, the hypothesis of micro-founded equilibrium (i.e., based on methodological individualism) and the absence of interaction between heterogeneous agents, which as we will see are the determinants of the distribution, imposes a straightjacket on the mainstream box of tools which inhibits it from any application on distribution. The interaction among agents can be identified with the *causa causans* of the nonlinearity that originates the distribution as an emergent phenomenon.

The laws of thermodynamics, among the most certain laws we have in physics, are the result of chance: random behavior determined by the interaction of billions of billions of molecules resulting in macroscopic regularity. This is what happens to the distribution of income and wealth, but in economics molecules are atoms in society and their dimensions are heterogeneous, so that there is the effect of St. Matthew (Mt 13:12): "Whoever has will be given more, and they will have an abundance. Whoever does not have, even what they have will be taken from them."[1] The heterogeneous interacting agents distribute themselves in a distribution characterized by a statistical equilibrium because of a self-organizing process with some technological and institutional constraints.

A natural starting point in this area of enquiry was the observation that the number of persons in a population whose incomes exceed x is often well approximated by $Cx^{-\alpha}$, for some real C and some positive α, as Pareto argued over 100 years ago. However, theoretical and empirical work rapidly pointed to the fact that it is only in the upper tail of the income distribution that a Pareto-like behavior can be expected, while the bulk of the income, the 95 % or so of the population, is governed by a completely different law. Therefore, many recent papers within this literature have sought to characterize the distribution of income by a mixture of known statistical distributions, even if there is a dispute about what these distributions are: indeed, while it seems to be generally acknowledged that the top 1–5 % of incomes follows the Pareto law, an exact and unequivocal characterization of the low to medium income region of the distribution is still evasive. Some scholars claim that this is lognormal; others maintain that the distribution of personal income for the majority of the population should follow the exponential law. In this book we address the issue of data analysis related to the size distribution of income by adopting a statistical mechanics approach introduced by Kaniadakis (2001, 2002, 2005), based on the one-parameter generalization of the exponential function. As we will show, our cumulative distribution function can describe the whole spectrum of the size distribution of income, ranging from the low region to the middle region, and up to the power-law tail, pointing in this way toward a unified approach to the problem.

The parameter k interacts with the layers one by one on the space of probabilities; k is a characteristic parameter, a something intrinsic element, a sort of "specific weight" of an element. All variables characterized by that parameter will be described by the same distribution; only the parameters of scale and form vary, because the data are different samples which come from different populations. In short, the k-distribution outperforms all other distributions: idiosyncratic characteristics of the data, as well as economic policy, influence constraints by changing the position of the distribution, but not its shape.

We wish to thank Giorgio Kaniadakis, the *father* of the k-distribution; Simone Landini, who wrote Appendix A and explained to us what we were writing, Costantino Tsallis, Jean-Paul Fitoussi, and Joe Stiglitz. We are also grateful to Stefano Spalletti for bringing to our attention some relevant contributions from the history of economic thought about the scope and methods of economics. Endless

[1] Endless wealth accumulation is well represented by the following anecdote: the not any more Cavalier B. once asked his 5-year-old son: "How old are you?"; at his reply he said: "Well, at your age I was 12."

discussions with our group in Ancona, in particular Ruggero Grilli, Antonio Palestrini, Luca Riccetti, and Alberto Russo, have been very stimulating. They survived brilliantly the meetings in which this book was discussed. Hopefully, the readers will too.

Macerata
Ancona
October 2015

Fabio Clementi
Mauro Gallegati

Contents

List of Figures

List of Tables

Chapter 1
Introduction

So distribution should undo excess, and each man have enough.
William Shakespeare

Abstract In this chapter, we discuss the motivations and goals of the research laid out throughout the book. The recent revival of interest in the study of income/wealth distribution is considered key to our analysis. Also, it is argued that the dominant paradigm in modern mainstream macroeconomics, the representative-agent general equilibrium framework, is badly equipped to cope with (re-)distributional issues. Instead, we emphasize that socio-economic systems are complex systems composed of heterogeneous interacting agents whose equilibrium is not individual, but statistical, meaning the state in which the distribution of agents' resources corresponds to the maximum probability among all possible distributions, i.e. to the "maximum entropy" probability distribution.

Keywords Income and wealth distributions · Macroeconomic modeling · Complexity · Heterogeneity · Interaction · Statistical equilibrium · Maximum entropy

1.1 The Revived Interest in the Problems of Income and Wealth Distribution

Income distribution has a long and rich tradition, from Ricardo and John Stuart Mill through Pareto, Clark and Obson, to Sen and Atkinson and, more recently, Stiglitz and Piketty.[1]

[1]The literature of the 1950s had identified as stylized facts: the constancy of the distribution shares of the aggregate income and the Kuznets curve. Recently, Stiglitz (2015b, c, d, e) has proposed others, which are truer for some countries than others and with only a few exceptions: (*i*) growing inequality in both wages and capital income (wealth), and growing inequality overall; (*ii*) wealth is more unequally distributed than wages; (*iii*) average wages have stagnated, even though productivity

F. Clementi and M. Gallegati, *The Distribution of Income and Wealth*,
New Economic Windows, DOI 10.1007/978-3-319-27410-2_1

The most impressive element of economic history is, undoubtedly, the exponential growth of income and population coinciding with the advent of the industrial revolution. Centuries of struggle for surviving were taken over by a kind of golden age due to technological innovations. Yet a so revolutionary phenomenon has its contradictions. On the one hand, there is a growing inequality between countries, between those who participate the economic development and others. Per-capita income doubled every 30 years, but the inequality does not change. At the same time span, GDP growth ratio between developed and underdeveloped countries was around 100. The curve of income and wealth distribution shifts up, not changing much of its shape.

If Smith takes care of the first problem, the interest Ricardo turns instead to is the domestic distribution, between rentiers, workers and capitalists. To do it, he needs a theoretical framework with more social groups (or classes). Fifty years later, the advent of the neoclassical economics buried such a research because it lacked all the tools for.[2]

During the two decades prior to the beginning of the global economic crisis, the gap between rich and poor grew in many of the G7 countries. From the late eighties to mid-nineties, the increase in income inequality was particularly accentuated in the United Kingdom, the United States and Canada. However, those years marked a growing gap between rich and poor, also—for the first time—in countries with traditionally low income inequality, such as Germany. In that same period only France and Japan did not register any increase in the levels of inequality.

In emerging countries, economic growth plays a decisive role in reducing poverty, but at the same time income distribution becomes more concentrated. Among the BRICS, only Brazil manages to substantially reduce income inequality—even if it is still high by world standards—while, in the same period, inequality in Russia, China, India and South Africa intensifies.

Italy is, among the OECD countries, one of those with the greatest inequalities in income distribution and, more importantly, wealth. And this occurs while the extremes grow further apart (the rich becoming richer and the poor poorer) and wealth begins to shift to the wallets of the elderly population at the expense of the younger ones. The Italian situation was much less unequal in the 1960s and early 1970s (even in 1992 the Gini coefficient was about 0.27).

Between the 1980s and the first decade of the millennium, the continental European countries maintained a practically constant index of inequality over time and between 0.23 of Norway and 0.34 of Greece, values well below the Italian one (0.32); the Nordic countries show a growing trend of inequality, although still characterized

(Footnote 1 continued)
has increased, so the share of capital has increased; (*iv*) significant increases in the wealth-to-income ratio; (*v*) the return to capital has not declined, even though the wealth-to-income ratio has increased. Of course, these are not the only stylized facts that need to be explained. There is a large literature trying to explain the shape of the income and wealth distributions, e.g. why the upper tails of the distributions are Pareto (fat) and why at lower levels of income and wealth the distributions are better described by alternative functional forms. This will become clearer later in the book.

[2]It is quite paradigmatic the fact that the Pareto efficiency of a market does not say anything about the distribution of income.

by low values (Sweden, 0.24; Finland, 0.26), while market economies (among which the United States, the United Kingdom and Australia) have more pronounced inequalities.

The series of the last decade shows that the economic crisis, causing high unemployment aggravated by *manoeuvres* of fiscal austerity as useless as they are harmful, has inequitably affected income distribution and determined a trend reversal: while in 2004 the Italian index is 0.33, and then gradually decreases to 0.31, starting from 2009, in the full-blown crisis, it begins to grow again.

According to Stiglitz (2012), in a vicious circle, greater inequality has led to less equality of opportunity, and therefore to greater inequality. Growing inequality jeopardizes support for collective action, which is what guarantees that everyone can fully develop his/her full potential as a result, for example, of investment in public education. Also, inequality fosters instability, which in turn gives rise to more inequality. Relocating money to high-income individuals, who in proportion consume a smaller slice of their earnings than the poor, reduces consumption and the multiplier (richer people save 15–25 % of their income, while the poor spend nearly all of it).[3] The result is that until something happens, like for example an increase in investments or exports, the total demand will be lower than what the economy would be able to meet, i.e. there will be unemployment. In the nineties that something was the technology bubble; in the first decade of the twenty-first century there was the housing bubble. Now the only solution is to resort to a redistribution policy and/or public spending.

Unemployment today can be attributed to a shortage arising from aggregate demand (total demand for goods and services in the economy by consumers, firms, government and exporters); in a certain sense, the entire current deficit in aggregate demand is due to phenomena of extreme inequality.

High inequality has important economic consequences: in general, the propensity to consume decreases with increasing income (and wealth). In this way, increasing inequality can lead to a reduction in aggregate demand. Credit consumption, through which savings of the rich finance consumption (liabilities of the poor), could support the insufficient demand. But debts accumulate and generate a massive redistribution of income from non-subscribers of debt securities to the subscribers (the two categories often coincide with the rich and the poor). Furthermore, financial markets provide great profit-making opportunities for those who have capital to invest. The high financial returns due to increasingly complex products can attract a lot of capital and the chances of inflating a bubble (which sooner or later bursts) are high. Financial expansion is, however, a phenomenon which is purely functional to capitalist development in a period of deregulation and rising inequality, because it allows to recycle profits from rising rates of exploitation. For these reasons, a *manoeuvre* of economic policy would be needed to increase taxes applied to 1 % of the population who for years has accumulated enormous wealth. In this way there would be resources to boost the economy by investing primarily in education, research and environment, while reducing inequality should stimulate a recovery in consumption.

---

[3]For the analytic of the inequality-adjusted multiplier see Gallegati et al. (2015).

The financial crisis of 2007–2008 and the Great Recession that followed left a large number of people in difficulty. Half a decade later, one European out of eight and one American out of eleven is unemployed. At the end of 2012, most of the savings of people who had lost their jobs three-four years previously had been spent.

The ominous roots of the economic boom of the previous decade now appear clearly visible. Although there has been growth in GDP, the majority of citizens see their standard of living fall compared to the previous ten years: economy has worked well only for the top layer of the population.

The history of income distribution inequality in recent years can be summarized as follows: the rich have become richer and, while the poor are poorer and more numerous, the middle class is disappearing. Over the past three decades, the pay of American low-wage workers grew by 15 %, that of Europeans by 7 %, while the top 1 % saw their income increase by almost 150 % (60 % in Europe).

Meanwhile, changes in wealth distribution are even more dramatic. Since much of the wealth of the middle and lower classes is established on the value (fictitious, since it's based on the real estate bubble) of their houses, when the bubble bursts, although everyone loses a bit of wealth, things are much worse for the poor. Even the rich lose a portion of their wealth, but in 2007 the richest 1 % of the richest American families had a net worth 225 times greater than the average American (130 in Europe), almost double the value in 1962 or in 1983.

Given the disparity of wealth, it's not surprising that those at the top of the pyramid have the lion's share of income growth in 2007: a little less than 60 % of that income goes to the top 1 %.

Economic growth in recent years has benefited the wealthier classes of the population. Moreover, countries with higher inequality are also likely to have an increase in both relative poverty and risk of poverty, along with a significant decrease in social mobility. The relationship between economic growth and income inequality is somewhat controversial. After the Second World War, for example, the process of economic growth coincides with a gradual reduction of inequalities that may have acted as a stimulus to aggregate demand through the multiplier. This is foreseen by the so-called Kuznets curve, according to which with an increase of income inequality first increases and then decreases, with a bell-shaped trend. What we are witnessing today is more like an "N" trend, with an increase in inequality in the more advanced stages of growth. From the seventies on, in fact, inequalities begin to grow again, and accumulated wealth has largely turned into a financial investment, where speculative bubbles promise yields that are far higher than those offered by real investments, the profits of which are limited by the increase in productivity and a decline in demand.

The causes that may lead to an increase in inequality are obviously numerous and we do not deal with them here.

"Inequality [. . .] is not just about the size of wallets" (Therborn 2013, p. 1). In most OECD countries, for example, over the last 25 years the ratio between the incomes earned by the richest 10 % and the poorest 10 % increased from 7 to 9. If the three richest people in the world own more financial assets of the 48 poorest nations, or the 85 richest individuals have a wealth equivalent to the 3.5 billion poor people, or the 1 % of the super-rich have more than 50 % of global wealth, we question not only whether there is any justification, but if the system is sustainable.

Fitoussi (2013) argues that an excessive level of inequality causes two effects: the unsustainability of a democratic framework, as the *élite* of the richest controls the media and there is no longer an economic policy other than the one dictated by the interests of the *élite* itself; unsustainability of human (education, health, safety) and social capital (security of the future, labor inclusion).

Since social capital, as well as environment, cannot be measured by the GDP, if we wanted to take into account the effect of trust on the entire economy considering precisely its social capital, we would have to change the system of measurement. An economy with more social capital is also more productive, just like an economy with more human or physical capital. Social capital is a broad concept which includes those factors that contribute to good governance in the public and private sectors. But the idea of trust is at the basis of any notion of social capital; people need to feel sure that they will be treated well, with dignity and equitably. And then they reciprocate. Social capital is the glue that holds societies together. If individuals believe that the economic and political system is unfair, the glue is unable to carry out its work and societies do not function well.

The government plays a dual role regarding the current level of inequality: it is partly responsible for the inequality in the distribution of gross (pre-tax) income and has reduced its role in correcting these disparities through progressive taxation and public spending policies. When the rich get richer they have more to lose regarding the attempts to limit their rent-seeking activities and to encourage a redistribution of income in order to create a more equitable economy. Also, they have more resources to resist these attempts. It could seem strange that, with increasing inequality, we have done very little to diminish the impact, even though, after all, it is what everyone would have expected. This is a trend which, no doubt, can be found more or less all over the world: the more an egalitarian society works hard to maintain its social cohesion, the more a society with strong inequalities will undertake policies (both public and private) to promote the persistence of inequality itself.

If we continued to think that by simply making the GDP grow everyone would automatically benefit, we would be wrong. Even if the economy produced more goods and services, if, year after year, most of today's citizens had incomes that were increasingly lower or higher, but at the expense of future generations, our economy would not be working well.

To conclude.

According to Hickel (2012), on the basis of the data presented by Milanovic (2002), the 358 richest people in the world have a wealth equal to that of the poorest 45 % of the global population. If we consider data for the three richest individuals in the world we are looking at a wealth that corresponds to that of the least developed countries put together, about 600 million people. Put simply, the richest 1 % of individuals holds about 40 % of the world's wealth; the poorest 50 % of the world population holds only 1 % of the total wealth. Is there any hope? The program of political and economic reforms argued by Stiglitz (2012) assumes that, while market forces play a role in creating the current level of inequality, their structure is, ultimately, a policy reform. We are (at least theoretically) able to remodel these market forces in order to promote greater equality, or at least a greater opportunity of equality.

1.2 Re-Incorporating Distributional Issues into the Main Body of Economic Analysis

Measurement of income inequality for evaluating social welfare is of particular interest to economics. Since the size distribution of income serves as a base for any operational inequality measure, correct specification of the income distribution shape is of great importance.

The study of the income size distribution has a long history. Pareto (1895, 1896, 1897a, b) was apparently responsible for the first attempt at defining a general "law" that tried to explain the regularities of observed distributions. However, rapidly accruing evidence showed that the Pareto distribution accurately models only high levels of income, but does a poor job in describing the middle and lower end of the distributions. Subsequently, many probability density functions have been proposed in the literature that are suitable for describing the size distribution of income amongst the population as a whole—see e.g. the survey contained in Kleiber and Kotz (2003). Fitting of parametric functional forms has also been common for the distribution of wealth.

This book is concerned with identifying a distribution that best describes the actual data on income and wealth. It is not for the "measurement without theory", because, as we shall see in this section, to accomplish this task involves leaving the toolbox of mainstream economics. Economy can be regarded as a complex system composed of heterogeneous interacting agents whose equilibrium is not individual, but statistical, meaning the state in which the distribution of agents' characteristics corresponds to the maximum probability among all possible distributions, i.e. to the "maximum entropy".

Economic systems are composed by agents whose actions cannot be considered in isolation, as if they were Robinson Crusoe. Their behaviour is influenced by the state of others. The outcome of this process of interaction is not just the aggregation of individual behaviour, but far richer because the system is complex (Anderson 1972). Some aspects of complex systems exhibit large-scale regularities, like the emergence of income and wealth distributions with heavy (Paretian) tails.[4]

[4]Yule (1925) showed that a heterogeneous, heavy-tailed distribution can emerge out of a simple stochastic process. He referred to a process of biological evolution. The Yule model is stochastically multiplicative, in the sense that the probability that a new species belongs to a given genus is proportional to the number of species the genus already contains. In other words and context, this model can explain the idea of the "rich get richer" effect, which determines the power-law behaviour. Statistical physics has also revealed that the laws of power, to which class belongs the Pareto law, are the signature of the "self-organized criticality" (Bak 1996), i.e. the mechanism opposed to "top-down" general equilibrium economics.

As a result of their interactions, heterogeneous agents can change their status, *ergo* the system configuration of agents—i.e. the way they are distributed on the different possible states of a space—can mutate. This implies that each state of the state space will be otherwise populated at any given time but also at different times.

When appropriate conditions occur that insist on the system and affect its constituents, one can reach a configuration when a stationary state is reached, that is the numerosity of the agents in different states does not change but the individual agent may change his status. At the same time there is an individual disequilibrium and aggregate balance.[5]

The observed distribution can represent a balance of the system, although its constituents are not because they continue to change positions (e.g. some rich become poor, and vice versa). Exchanges in this case are compensatory: if one moves from B to enter the state A (rich $\rightarrow$ poor), then one comes from the state A to enter B (poor $\rightarrow$ rich). If this mechanism is indefinitely repeated in time, then we are facing a situation that, by analogy with the economic jargon but with little accuracy, one may define as a *steady state*, i.e.: the distribution was carried to the stationary state because the density of positions does not change even if the constituents of the system continue to change positions. This is the basic idea of *statistical equilibrium* (Aoki 1996, 2002; Aoki and Yoshikawa 2007; Foley 1994).

In this context, one cannot use the representative agent (RA) framework, or even a variation of it, because it has a lot of logical contradictions and is completely unfitted to describe a complex system.

First of all, there is the problem of how to fit data with a lot of heterogeneity within the RA framework. Income and wealth distributions typically have a large-size domain spanning several orders of magnitude.

From a political economy side, there is the problem of redistribution, which is quite impossible within the RA framework.

Moreover, the introduction of several different RAs can deal with the problem of absolute poverty, but not with the relative one which endogenously changes because of the *mean-field* problem. The problem is the same general equilibrium economics has with endogeneity in expectation formation, well explained in the rational expectations literature: one needs to know the true model to obtain coherent expectations. In the case of income and wealth distributions this tantamount to say that one can use heterogeneity as different RAs if he knows the distribution of the agents, i.e. a fallacy of *petitio principii* (begging question) or *circulus in probandum* (circular demonstration) in which one is assuming what should be proved.

The complexity typical of the economic system requires a statistical equilibrium approach: the possibility of equilibrium at the macro level in the presence of a multitude of situations of non-equilibrium at the microscopic level, where data show agents continuously interchange their positions.

The all is not the sum of the parts because of heterogeneity and interaction: they are two sides of the same coin; one implies the other and vice versa. The sum of all

---

[5]The old *Gattopardo* adage: "everything changes [at the micro level] so that nothing changes [at the macro level]".

the numbers (density) of the different states of occupation affects the volume (unit) of the total number of agents, but this is only the numerical result of a mechanism which is described microscopically by functions only exceptionally linear, and the sum of the functions of the parts is not the aggregate function (Landini and Gallegati 2014).

The heterogeneity implies interaction. And it involves non-linearity and externalities which are destructive to the mainstream model. The heterogeneity that after the Great Recession has been massively introduced in the dynamic stochastic general equilibrium (DSGE) models should carefully avoid interaction. There may be poor and rich agents, but it should be forever. One can solve the DSGE models with heterogeneous agents, but for each class of agents and then summing up the solutions. The only way to take into account of heterogeneity with equilibrium is to avoid the problems of coordination and strategy. Applied to the problem of distributions, this is equivalent to saying that we can study their shape already knowing the shape to be investigated.

We seek a way to infer distribution capable of connecting at the same time macro-equilibrium with micro-disequilibrium. Consider the distribution out of the steady state; it is characterized by an unknown generating mechanism. Therefore, we have to "read" it according to a probabilistic language, as it is done in statistical mechanics. We can infer a functional form of the distribution that underlies the system using the scarce information available, obviously under some conditions. For example, we could know some estimates of certain moments of a distribution and look for the "most accurate" distribution function, given the context of limited information we are. In this situation, it is natural to resort to the maximum entropy principle.

Each probability distribution is characterized not only by its moment generating function, but also by its probability/cumulant generating function and its characteristic function. In addition, a density is associated to a single measure of entropy, while from a measure of entropy can descend a family of density. Whatever the measure of entropy chosen, if the configuration alters, the value of entropy changes except in the case where the configuration becomes stationary, i.e. does not change any more. In this case the invariance of entropy means that a stationary point has been reached.

We consider the (pseudo-)measurement according to Shannon(-Boltzmann-Gibbs) entropy. This measure is concave, like every other entropy (hidden legacy of the Second Law of Thermodynamics, where the entropy is strictly defined from the phenomenological point of view), and, if it exists, must reach a maximum as a stationary point: that point is the maximum value of the concave function. Since no income or wealth measure can assume undefined values, we assume that his range of variation is necessarily compact, limited and closed, even though there are not always well known extremes. We wonder what is the configuration that maximizes the Shannon entropy.

To formalize, let $W(i) = w$ be the realization of individual income and $p(h)$ denote the probability that the ith income falls within the hth class. The estimator of the expected value for the sample of incomes W is $E[W] = p(1)w(1) + \cdots + p(h)w(h) + \cdots + p(H)w(H)$. Suppose you know the estimate of the average income, m, and put $E[W] = m$. The sequence $w(1), \ldots, w(H)$ can be predeterminably known according to some criterion (for example, for $H = 2$ we could distinguish conventionally between rich and poor, or give a graduation more refined if it were $H > 2$). One can wonder: what is the estimator of $p(h)$ that builds the distribution $p(1), \ldots, p(H)$ more consistent with the information known? In other words: what is the distribution of W more consistent with the information known about $E[W] = m$ according to an optimality criterion that maximizes the Shannon entropy given by $S = -p(1)\ln[p(1)] - \cdots - p(h)\ln[p(h)] - \cdots - p(H)\ln[p(H)]$?

By employing the principle of the Lagrange multipliers, the solution to the question is a function of probability, $p(w)$, in the form of Gibbs distribution, possibly modulated by a few parameters. $p(w)$ is the distribution that maximizes the entropy under the given constraints, so that it is the most likely specification of the data. It also changes over time, i.e. $p(w,t) = n(w,t) = N(w,t)/N(t)$, returning an inference that meets a criterion of optimality (the entropy maximization) consistently with a given macro-state possibly mutable over time, being $m(t)$ known. Therefore, one does not care who enters or leaves the state w in time. Rather, one wants to find the situation where $p(w,t) = p(w)$ consistently with $m(t)$, i.e. when there is a steady-state distribution, a state known as equilibrium statistical system.

1.3 Aim and Contents of this Book

To recapitulate: the inequality changes over time, but its functional form does not. The distribution of income or wealth we observe today is roughly similar to those of the Egypt's Pharaoh Akhenaten (Abul-Magd 2002), or the medieval Hungarian economy (Hegyi et al. 2007), or Italy of the late nineteenth century (Pareto 1895, 1896, 1897a, b). Thus the research questions: is there a law governing the distribution, although the coefficients of it may change? What form assumes the distribution (not what determines it) and which function best describes it?

This book is devoted to presenting a systematic overview of cutting-edge research in the field of parametric modelling dealing with income and wealth distributions. While reviewing the state-of-the-art of the field, we will provide a thorough discussion of parametric models belonging to the "κ-generalized" family, a new and fruitful set of statistical models for the size distributions of income and wealth developed by the authors over several years of collaborative and multidisciplinary research. It will be given knowledge of their basic statistical properties and evidence of their empirical plausibility, with the explicit aim of demonstrating that they are natural and sometimes superior alternatives to other widely used parametric models for the distributions of income and wealth.

Given the renewed focus on distributional issues, and motivated by the feeling that problems of income and wealth distribution should be given the conceptual and methodological attention they merit from the economics profession, we hope to convince readers that this class of distributions deserves more attention than it has received until very recently.

Chapter 2
The Parametric Approach to Income and Wealth Distributional Analysis

Essentially, all models are wrong, but some are useful.
George Edward Pelham Box

Abstract This chapter provides a brief overview of the main advantages associated with using parametric models of income/wealth distribution. It also instructs the reader about the numerous statistical models of income/wealth distribution that have been proposed in both the statistical and economic literature for over 100 years since Pareto's breakthrough contribution.

Keywords Personal income, wealth and their distributions · Parametric modeling · Literature review

2.1 The Idea of a Parametric Model for Income and Wealth Distributions

The analysis of a probability density function is a powerful tool to describe the full distribution of a variable of interest and one which reveals much more information of its several properties compared to standard descriptive statistics, such as the mean, variance, skewness or kurtosis, summarizing each limited properties of the distribution on single values (Cowell and Flachaire 2015). For instance, the analysis of the shape of the income/wealth distribution provides a picture from which at least three important distributional features can be observed simultaneously (Cowell et al. 1996): income/wealth levels and changes in the location of the distribution as a whole; income/wealth inequality and changes in the spread of the distribution; clumping and polarization as well as changes in patterns of clustering at different modes.

In practice, the shape of the density function is unknown and has to be estimated from data on hand. For a long time, the main estimation method was parametric,

F. Clementi and M. Gallegati, *The Distribution of Income and Wealth*,
New Economic Windows, DOI 10.1007/978-3-319-27410-2_2

which requires the choice of a functional form a priori. In the last decades, the availability of powerful computer resources gave rise to various non-parametric methods of density estimation, the most popular being the kernel density estimators which are often used in empirical studies now. Their main attraction consists in the possibility to relax the specific assumptions underlying parametric estimation method, as no distributional form is imposed a priori. However, with small data sets kernel density interpolation might result in imprecise estimates; also, the accuracy of this method typically depends to a large extent on the bandwidth used in the kernel functions.[1] These inaccuracies of model-free density estimation may be reduced using the parametric technique, whose alleged "old-fashionedness"—given its 100-year-plus history as opposed to the relatively young non-parametric methods—has been earnestly defended, among others, by Piketty (2000, p. 462) in his book review of Champernowne and Cowell's (1998) *Economic Inequality and Income Distribution*:

> One criticism that could be addressed to the book is that some readers might find it a little bit old-fashioned. For instance, the authors refer frequently to Pareto coefficients and the way such coefficients can be used to measure income inequality or to fit a Pareto distribution. Many young readers have probably never heard of such techniques, and most of them will probably never use them: at the age of SAS and Stata, one simply needs to push a button on a computer to compute the deciles, centiles and various inequality measures associated with a given income data base. In my view, however, such a criticism would be misplaced: the point is that individual-level databases on incomes have been available only for the past few decades and for the most developed countries, so that these 'old-fashioned' techniques are still very useful. In addition, it seems to me that it is very useful to have a solid background about past economists' work on income distribution (instead of believing that serious research started in 1980 or in 1990, which today's economists frequently tend to assume), and this book provides such a background.

Of course, using parametric models in the analysis of the distribution of income and wealth is not without advantages. For instance, fitting parametric models allows one to represent the entire distribution through a small number of estimated parameters (Brachmann et al. 1996). The estimated parameters may then be used to reconstruct the entire distribution if, for example, income or wealth data released in future are published in grouped form (Hajargasht et al. 2012) or if available micro-data are censored to bounds (i.e. bottom and/or top-coded; see Feng et al. 2006, and Burkhauser et al. 2011). Reconstruction of the overall distribution can also be achieved with the help of a reliable parametric model when only empirical estimates of poverty and/or inequality measures are available—such as those published by the

[1] In standard kernel method the bandwidth remains constant at all points where the distribution is estimated. This constraint can be particularly onerous when the concentration of data is markedly heterogeneous in the sample. Hence, there would be advantages from using a narrower bandwidth in the dense part (the middle) of the distribution and wider ones in the more sparse tails—as in "adaptive" kernel estimation (Van Kerm 2003)—especially in the cases of heavy-tailed income and wealth distributions. Greater modeling flexibility can also be achieved by means of finite mixture densities, defined as convex combinations of two or more parametric densities. The separate analysis of the components of the mixtures and of the relative importance of these components over time are the main advantages over a non-parametric approach, as they allow capturing the effect of unobserved heterogeneity (Cowell and Flachaire 2015).

World Bank, OECD, Eurostat or other statistical agencies—and direct access to the underlying micro-data is not possible (Graf and Nedyalkova 2014). Furthermore, as maintained by Cowell (2000, p. 145):

> some standard functional forms claim attention, not only for their suitability in modeling some features of many empirical [. . .] distributions but also because of their role as equilibrium distributions in economic processes.

The parameters of theoretical models often possess also economic interpretation, which allows to gain insights about the causes of the evolution of the distribution over time or interpret the differences between income/wealth distributions across countries. Moreover, once a given parametric model is fitted to a data set, one can straightforwardly compute inequality and other welfare indicators as analytical functions of the parameters of the model. It is also possible to use estimated parameters to perform stochastic dominance testing for inequality and welfare differences between distributions (Kleiber and Kotz 2003) as well as to empirical model the impact of macroeconomic conditions (e.g. GDP growth, unemployment and inflation rates, etc.) on the evolution of the personal income or wealth distribution—see Jäntti and Jenkins (2010) and references therein. Finally, estimated parameters may be made a function of covariates summarizing personal characteristics; this allows distributional shape to vary with population characteristics and provides a route to decomposition analysis of the sources of trends in income/wealth distributions over time or differences between countries (Betti et al. 2008; Biewen and Jenkins 2005; Quintano and D'Agostino 2006).

2.2 Brief History of the Models for Studying Income and Wealth Distributions

The interest in finding parametric models for the size distributions of income and wealth has a 100-year-plus history that dates back to the work of Pareto (1895), who was the first to propose a functional form for approximating the observed distributions.[2] The Pareto model—usually referred to as the (strong) "Pareto law"—is a two-parameter distribution with a power-law density on the support $[x_0, +\infty)$; it was found to accurately model high levels of income/wealth, but did a poor job in describing the lower-middle range of the distribution. Based on Pareto's economic foundations, and on the stochastic foundations afterward developed by other authors (e.g. Mandelbrot 1960 and Ord 1975), the Pareto law is now overwhelmingly considered as the parametric model of the rich.[3]

[2]There exists a huge amount of literature on parametric models for the size distributions of income and wealth. Here we limit ourselves to consider the most frequently cited contributions in the area. For the interested reader, a comprehensive survey can be found in Kleiber and Kotz (2003).

[3]In his pioneering contributions at the end of the nineteenth century, Pareto (1896, 1897a, b) suggested two variants of his distribution, occasionally called the three-parameter Pareto distributions. These further Pareto distributions, however, have not been used much in empirical economic studies.

The use of other density functions was later advocated by Gibrat (1931), who provided a theoretical basis (the "law of proportionate effect") for the two-parameter lognormal to be considered as a model for the size distributions of income and wealth. The lognormal distribution was further explored by Aitchison and Brown (1954, 1957) and Lebergott (1959). Another two-parameter statistical model, the gamma, was introduced by Salem and Mount (1974). Later evidence, however, showed that while these two-parameter models fit the data relatively well in the middle range, they tend to exaggerate the skewness and perform poorly in the upper end of the empirical distributions (McDonald and Ransom 1979). Also, they do not allow for intersecting Lorenz curves sometimes observed with income/wealth data (Kleiber 2008b).

Better fits as well as intersecting Lorenz curves were obtained using three-parameter models. Among these, the generalized gamma (Atoda et al. 1988; Esteban 1986; Kloek and van Dijk 1978; Taillie 1981), Singh-Maddala (1976) and Dagum type I (Dagum 1977) distributions are the most popular. These models possess the important property of the "weak Pareto law" (Mandelbrot 1960), that is they converge in distribution to the Pareto model for sufficiently large values of income and wealth. This important convergence property, motivated by the undisputed acceptability of the Pareto distribution as the model of the rich and the extremely rich, is strengthened by their power to describe with similar accuracy the remaining parts of actual distributions, i.e. the lower and middle ranges of income and wealth.

Models with more than three parameters have also been suggested in the size distributions literature for fitting income and wealth data. As a matter of example, the generalized beta distribution of the second kind (hereafter referred to as GB2) is a four-parameter distribution introduced by McDonald (1984) which has not only been very successful in fitting the data, but also includes some of the previously mentioned two- and three-parameter models as special or limiting cases. Other four-parameter models are the generalized beta distribution of the first kind (McDonald 1984), shortly referred to as GB1, and the double Pareto-lognormal distribution (Reed and Jorgensen 2004). Both of these models have proven to work remarkably well in fitting the data, although the former is sometimes outperformed by the GB2 and the three-parameter Dagum type I and Singh-Maddala distributions (Bandourian et al. 2003; Bordley et al. 1996).

McDonald and Xu (1995a, b) also developed the five-parameter generalized beta distribution family, which includes the GB1 and GB2 as special cases—of course along with all of the two- and three-parameter distributions nested within them. In turn, the double Pareto-lognormal distribution was generalized to a five-parameter family of distributions and was called the generalized double Pareto-lognormal distribution (Reed 2007). However, closed-form expressions of the probability density and/or cumulative distribution functions do not always exist for these "super" models, which makes fitting them to data computationally difficult and slow because of the need to resort to numerical techniques (McDonald and Ransom 2008; Reed and Wu 2008). Furthermore, and of considerable importance, one might question about the usefulness and meaning of five-parameter models to approximate income and wealth distributions and could consider them to be just curious theoretical generalizations (Kleiber and Kotz 2003).

Recently, Clementi et al. (2007, 2008, 2009) developed the three-parameter κ-generalized distribution, a non-Gaussian distribution with power-law tails having its roots in the framework of statistical mechanics. This distribution is a better fit for some countries than the Singh-Maddala, Dagum type I and GB2 models (Clementi et al. 2010, 2012a) and tends to yield better estimates of income inequality even when the goodness-of-fit is inferior to the existing personal income distributions (Okamoto 2012a, b). The κ-generalized distribution was also successfully used in a three-component mixture model for handling the distinctive features of wealth survey data (Clementi et al. 2012b) and later extended to four-parameter models for the size distributions of income and consumption (Okamoto 2013). Since much of this book is a systematic exposition of all that appeared in the recent literature and was known to us about the κ-generalized distribution and its extensions, conceptualization and novel empirical research for this "family" of κ-generalized models will be discussed in depth in subsequent chapters.

Chapter 3
The κ-Generalized Distribution

> *You have to learn the rules of the game. And then you have to play better than anyone else.*
>
> Albert Einstein

Abstract This chapter introduces the κ-generalized distribution and its interrelations with other statistical distributions. It also provides the basic statistical properties and inferential aspects of this parametric model as well as aspects that are of special interest in the income distribution field, including the Lorenz curve and inequality measures. The chapter ends with many up-to-date applications to income data, showing how accurately the κ-generalized distribution describes the data-generating process and the associated level of inequality.

Keywords Income size distribution · Income inequality · Parametric modeling · κ-generalized distribution

3.1 The Distribution and Its Properties

The new distribution presented here exploits recent advances on the use of κ-deformed exponential and logarithm functions introduced by Kaniadakis (2001, 2002, 2005) in the context of special relativity. Within this theoretical framework, the ordinary exponential function $\exp(x)$ deforms into the generalized exponential function $\exp_\kappa(x)$ given by

$$\exp_\kappa(x) = \left(\sqrt{1+\kappa^2x^2}+\kappa x\right)^{\frac{1}{\kappa}}, \quad x \in \mathbb{R}, \quad \kappa\,[0, 1)\,. \tag{3.1}$$

The deformed logarithmic function $\ln_\kappa(x)$ is defined as the inverse function of (3.1), namely $\ln_\kappa\left[\exp_\kappa(x)\right] = \exp_\kappa\left[\ln_\kappa(x)\right] = x$, and reads as

F. Clementi and M. Gallegati, *The Distribution of Income and Wealth*, New Economic Windows, DOI 10.1007/978-3-319-27410-2_3

$$\ln_{\kappa}(x) = \frac{x^{\kappa} - x^{-\kappa}}{2\kappa}, \quad x \in \mathbb{R}_{+}. \tag{3.2}$$

In the last few years the Kaniadakis deformed functions were adopted successfully to analyze also non-physical systems, including economic systems. In particular, the κ-deformation has been employed for studying differentiated product markets (Rajaonarison 2008; Rajaonarison et al. 2005) and for modeling the size distribution of income (Clementi et al. 2007, 2008, 2009, 2010, 2012a). As we will see, using such deformed functions in the latter context is attractive because they allow one to statistically describe the whole spectrum of incomes, ranging from the low region to the middle region, and up to the Pareto tail. For the interested reader, some further mathematical aspects of these functional relationships are reviewed shortly in Appendix A (see Sect. A.7.1).

3.1.1 Derivation

Given the mathematical formalism previously introduced, the κ-generalized statistical model can be derived from the maximum entropy principle, which offers an appealing way to select the best model without making any (independence or other) assumption on the data.[1] The "maximum entropy principle", largely due to Jaynes (1957a, b), states that of all distributions satisfying certain constraints one should select the one with largest entropy, which is a measure of uncertainty associated with a random variable. In this way it is guaranteed that a probability distribution does not incorporate any extraneous information other than that specified by the relevant constraints—typically moment constraints—which represent particular knowledge (evidence) about the random variable in question. Finding the maximum entropy distribution could be therefore considered as a general inference procedure, and indeed it was proposed initially by Jaynes (1957a, b) in this manner.[2]

[1] Virtually, all income distributions can be derived from the maximum entropy principle. For example, Ord et al. (1981) and Leipnik (1990) pointed out that several well-known income densities—such as Pareto, lognormal, gamma, Singh-Maddala, and GB2—might be selected if one uses a criterion of maximum entropy. Nonetheless, most of these densities are characterized by just two moment functions that take care of lower, middle and upper income levels at the same time. As it will become clear later in the chapter (see Sect. 3.2.3), these two moment functions are not enough to extract all the relevant information from the data, whereas adding one more moment function to the maximum entropy problem makes the specification under consideration more consistent with the amount of information (or uncertainty) present in the data, while achieving almost identical (when not definitely superior) goodness of fit.

[2] In Jaynes' (1957a, p. 623) words, "[...] in making inferences on the basis of partial information we must use that probability distribution which has maximum entropy subject to whatever is known. This is the only unbiased assignment we can make; to use any other would amount to arbitrary assumption of information which by hypothesis we do not have".

The maximum entropy principle has been applied in a great variety of fields including statistical mechanics, statistics, stock market analysis, queuing theory, image analysis and reliability estimation (see e.g. Kapur 1989 for a comprehensive account). It has also been adopted in the analysis of income distribution to rigorously describe the features of personal income data.[3] For instance, Wu (2003) introduced moment constraints into the maximum entropy problem sequentially until they failed to improve the model fit. For each set of moment constraints, the maximum entropy parameters were estimated and various model selection criteria calculated in order to find the best-fitting distribution. Based on this method, Wu (2003) was able to conclude that the best description of the 1999 US income distribution should include 8 to 12 moment constraints in the maximum entropy problem. A similar approach was taken by Wu and Perloff (2005), who applied the same technique to Chinese income data from 1985 to 2001. However, while extremely sophisticated, this methodology makes the work of linking the distributional features to economic interpretation very taxing, as some of the moment constraints are introduced without considering stylized facts and economic implications of the resulting density functions; furthermore, existence of high-order moment constraints can not always be ensured (Park and Bera 2010).

In the present section, constraints to the maximum entropy problem that represent distributional characteristics of income data more directly are introduced in a parsimonious and intuitive way. Following Jaynes (1978), the idea is to see what a comparison of the entropy in the actual data with the entropy based on the proposed statistical model reveals about which constraints should be included in the maximum entropy problem.[4] In particular, we consider in the sequel three moment constraints that take care of lower, middle and upper income levels at the same time; what we end up with is the three-parameter density function

$$f\left(x;\alpha,\beta,\kappa\right)=\frac{\alpha}{\beta}\left(\frac{x}{\beta}\right)^{\alpha-1}\frac{\exp_{\kappa}\left[-\left(x/\beta\right)^{\alpha}\right]}{\sqrt{1+\kappa^{2}\left(x/\beta\right)^{2\alpha}}},\quad x>0, \tag{3.3}$$

where $\alpha, \beta > 0$ and $\kappa \in [0, 1)$, which obeys the weak Pareto law and satisfy other empirical stylized facts of personal income data—such as possessing at least one interior mode and allowing for intersecting Lorenz curves.

[3]Broadly speaking, the use of the *entropy* concept in the analysis of income distribution is not new. Dalton (1920) “principle of population” paved the way of introducing a general measure of inequality that led to the notion of entropy-like function much earlier to the work of Shannon (1948) in information theory. This suggestion then found expression in the entropy-based measure of inequality proposed by Theil (1967), which naturally contributed to the development of a general information-theoretic approach to the measurement of inequality (Cowell 1980a, b; Cowell and Kuga 1981a, b).

[4]Comparing the entropy of the data to the entropy based on a particular distribution function makes it possible indeed to judge whether relevant moment constraints have been neglected. As Jaynes (1978, p. 47) puts it, “Any statistically significant departure from a maximum-entropy prediction then constitutes strong—and if persistent, conclusive—evidence of the existence of new constraints that were not taken into account in the calculation”.

More formally, let

$$S \equiv -\int_0^\infty f(x) \ln[f(x)] \, dx \tag{3.4}$$

be the Shannon (1948) entropy measure[5] to be maximized subject to the natural constraint that normalizes the density,

$$\int_0^\infty f(x) \, dx = 1, \tag{3.5}$$

and the three characterizing moments

$$\int_0^\infty \ln(x) f(x) \, dx = \ln(\beta) - \frac{1}{\alpha}\left[\gamma + \psi\left(\frac{1}{2\kappa}\right) + \ln(2\kappa) + \kappa\right], \tag{3.6}$$

$$\int_0^\infty \ln\left[1 + \kappa^2\left(\frac{x}{\beta}\right)^{2\alpha}\right] f(x) \, dx = 2\kappa - \psi\left(1 + \frac{1}{4\kappa}\right) + \psi\left(\frac{1}{2} + \frac{1}{4\kappa}\right), \tag{3.7}$$

$$\int_0^\infty \sinh^{-1}\left[-\kappa\left(\frac{x}{\beta}\right)^{\alpha}\right] f(x) \, dx = \kappa, \tag{3.8}$$

where the latter reproduces the deformation parameter κ and is introduced to take care of the thickness of the right-tail part of the distribution.

The constraint (3.6) is related to the sample geometric mean, and captures the distributional pattern of lower income levels. The arithmetic mean on the log-scale—or the logarithm of the geometric mean—is smaller than the arithmetic mean, and this makes an essential constraint for positively skewed random variables, as samples drawn from right- and (even more) heavy-tailed distributions exhibit values located on the right area very far from the mean value. In a sense, these values act like "outliers", and consequently strongly influence sample moments like the arithmetic mean, which is a classical measure of central tendency and one of the most commonly used constraints. Therefore, it is not rational to assume that the arithmetic mean, especially based on samples drawn from heavy-tailed distributions, is likely to be preserved. On the contrary, the function $\ln(x)$, which exhibits the convenient

[5]The Shannon entropy is not the only measure of uncertainty related to a random variable. In the literature, many different entropy measures have been proposed, most of which are generalizations of the Shannon one (for a summary of these measures, see e.g. Esteban and Morales 1995). Nevertheless, the Shannon entropy continues to be used in many works as the most well-justified measure of uncertainty for continuous random variables.

property for incomes to be defined only for positive values, eliminates the influence of these extreme values and offers a very robust measure that is more likely to be preserved than the estimated arithmetic mean.[6] It is not surprising, therefore, that most well-known income distributions are characterized by this moment condition.[7]

Finally, the constraint (3.7), introduced to take care of the middle and higher income levels, is a generalization of the classical moments based on the logarithmic and power functions, which naturally leads to power-law type distributions (Papalexiou and Koutsoyiannis 2012). It exhibits all the desired properties, like those of the $\ln(x)$ function described above, and thus is suitable for positive, highly varying and right-skewed random variables.

By applying the standard technique of constrained maximization, the Lagrangian of the continuous maximum entropy problem (3.4)–(3.8) is given by

$$\mathcal{L} = S - \sum_{j=0}^{3} \lambda_j \left[\int_0^\infty \phi_j(x; \boldsymbol{\theta}) f(x) \, \mathrm{d}x - C_j \right], \tag{3.9}$$

where λ_j are multipliers. The expressions for the functions $\phi_j(x; \boldsymbol{\theta})$, involving the parameter vector $\boldsymbol{\theta}$, and the constants C_j can be recovered by matching terms in the set of constraints (3.5)–(3.8). In particular, the normalization constraint (3.5) corresponds to $j = 0$ by setting $\phi_0(x; \boldsymbol{\theta})$ and C_0 equal to 1.

The first-order condition

$$\frac{\partial \mathcal{L}}{\partial f(x)} = -\ln[f(x)] - (1 + \lambda_0) - \sum_{j=1}^{3} \lambda_j \phi_j(x; \boldsymbol{\theta}) = 0 \tag{3.10}$$

implies that the solution has the functional form

$$f(x; \boldsymbol{\theta}) = \frac{1}{\Omega(\boldsymbol{\theta})} \exp\left[-\sum_{j=1}^{3} \lambda_j \phi_j(x; \boldsymbol{\theta}) \right], \tag{3.11}$$

where

$$\Omega(\boldsymbol{\theta}) = \exp(1 + \lambda_0) = \int_\infty^0 \exp\left[-\sum_{j=1}^{3} \lambda_j \phi_j(x; \boldsymbol{\theta}) \right] \mathrm{d}x \tag{3.12}$$

is the partition function that normalizes the density and can be determined from Eq. (3.5).

[6]Essentially for this reason, the logarithmic transformation is probably the most common transformation used in econometrics, as it tends to normalize positively skewed data.

[7]For maximum entropy characterization of some common income densities, see for instance Park and Bera (2010).

In order to determine the values of the multipliers as a function of the parameters $\boldsymbol{\theta} = \{\alpha, \beta, \kappa\}$, we observe that for each j it holds (Kaniadakis 2009)

$$\lambda_j = \frac{\partial S}{\partial C_j}. \tag{3.13}$$

Hence, taking the logarithm of Equation (3.3), multiplying it by $[-f(x)]$ and integrating between 0 and ∞, we get

$$\begin{aligned} S &= \ln\left(\frac{\beta^\alpha}{\alpha}\right) - (\alpha - 1)\int_0^\infty \ln(x)\, f(x)\,\mathrm{d}x \\ &\quad + \frac{1}{2}\int_0^\infty \ln\left[1 + \kappa^2\left(\frac{x}{\beta}\right)^{2\alpha}\right] f(x)\,\mathrm{d}x \\ &\quad - \frac{1}{\kappa}\int_0^\infty \sinh^{-1}\left[-\kappa\left(\frac{x}{\beta}\right)^{\alpha}\right] f(x)\,\mathrm{d}x \\ &= \ln\left(\frac{\beta^\alpha}{\alpha}\right) - (\alpha - 1)\, C_1 + \frac{1}{2}C_2 - \frac{1}{\kappa}C_3, \end{aligned} \tag{3.14}$$

which, applying (3.13), produces

$$\lambda_1 = 1 - \alpha, \quad \lambda_2 = \frac{1}{2} \quad \text{and} \quad \lambda_3 = -\frac{1}{\kappa}. \tag{3.15}$$

Finally, by inserting these results into (3.12) and solving the integral yields $\Omega(\boldsymbol{\theta}) = \frac{\beta^\alpha}{\alpha}$, which can be substituted for together with (3.15) into (3.11) leaving us with the density function (3.3).[8]

3.1.2 *Definitions and Interrelations*

A random variable X is said to have a κ-generalized distribution, and we write $X \sim \kappa\text{-gen}(\alpha, \beta, \kappa)$, if it has a probability density function given by (3.3). Its cumulative distribution function can be expressed as

[8]A brief point of clarification is in order here: the maximum entropy approach is not a method of assigning probabilities from scratch. With no constraints apart from normalization, one gets a uniform distribution, which is lurking in the background as a "prior" that is assumed by the maximum entropy method. In other words, maximum entropy updates probabilities from a previous distribution, it does not generate them from scratch (e.g. Caticha 2012). For this reason, a more heuristic derivation of the κ-generalized density will be given in Appendix A, Sect. A.7.2, where it will be shown how this probability distribution naturally emerges within the field of κ-deformed analysis.

$$F(x; \alpha, \beta, \kappa) = 1 - \exp_\kappa \left[-(x/\beta)^\alpha \right]. \tag{3.16}$$

Figure 3.1 illustrates the effects on the shape of the distribution of different values of the parameters. The exponent α quantifies the curvature (shape) of the distribution, which is less (more) pronounced for lower (higher) values of the parameter. The constant β is a characteristic scale, since its value determines the scale of the probability distribution: if β is small, then the distribution will be more concentrated around the mode; if β is large, then it will be more spread out. Finally, the parameter κ measures the heaviness of the right tail: the larger (smaller) its magnitude, the fatter (thinner) the tail.

As $\kappa \to 0$, the distribution tends to the Weibull distribution;[9] it can be easily verified that

$$\lim_{\kappa \to 0} f(x; \alpha, \beta, \kappa) = \frac{\alpha}{\beta} \left(\frac{x}{\beta} \right)^{\alpha - 1} \exp\left[-(x/\beta)^\alpha \right] \tag{3.17}$$

and

$$\lim_{\kappa \to 0} F(x; \alpha, \beta, \kappa) = 1 - \exp\left[-(x/\beta)^\alpha \right]. \tag{3.18}$$

For $x \to 0^+$ the distribution behaves similarly to the Weibull model (3.17) and (3.18), whereas for large x it approaches a Pareto distribution with scale $\beta(2\kappa)^{-\frac{1}{\alpha}}$ and shape $\frac{\alpha}{\kappa}$, i.e.

$$f(x; \alpha, \beta, \kappa) \underset{x \to +\infty}{\sim} \frac{\frac{\alpha}{\kappa} \left[\beta (2\kappa)^{-\frac{1}{\alpha}} \right]^{\frac{\alpha}{\kappa}}}{x^{\frac{\alpha}{\kappa}+1}} \tag{3.19}$$

and

$$F(x; \alpha, \beta, \kappa) \underset{x \to +\infty}{\sim} 1 - \left[\frac{\beta (2\kappa)^{-\frac{1}{\alpha}}}{x} \right]^{\frac{\alpha}{\kappa}}, \tag{3.20}$$

thus satisfying the weak Pareto law (Mandelbrot 1960).[10]

Equation (3.16) implies that the quantile function is available in closed form

$$F^{-1}(u; \alpha, \beta, \kappa) = x_u = \beta \left[\ln_\kappa \left(\frac{1}{1-u} \right) \right]^{\frac{1}{\alpha}}, \quad 0 < u < 1, \tag{3.21}$$

[9]The Weibull distribution was used only sporadically as an income distribution. Some applications can be found in Bartels and van Metelen (1975), Bartels (1977), Espinguet and Terraza (1983), McDonald (1984), Atoda et al. (1988), Bordley et al. (1996), Brachmann et al. (1996) and Tachibanaki et al. (1997).

[10]Further generalizations of the Pareto law were introduced by Kakwani (1980), $\lim_{x \to +\infty} \frac{xf(x)}{1-F(x)} = \alpha$, and Esteban (1986), $\lim_{x \to +\infty} \left[1 + \frac{xf'(x)}{f(x)} \right] = -\alpha$. Since we have

$$\lim_{x \to +\infty} \frac{xf(x; \alpha, \beta, \kappa)}{1 - F(x; \alpha, \beta, \kappa)} = \frac{\alpha}{\kappa} \quad \text{and} \quad \lim_{x \to +\infty} \left[1 + \frac{xf'(x; \alpha, \beta, \kappa)}{f(x; \alpha, \beta, \kappa)} \right] = -\frac{\alpha}{\kappa},$$

the distribution also obeys these alternative versions of the weak Pareto law.

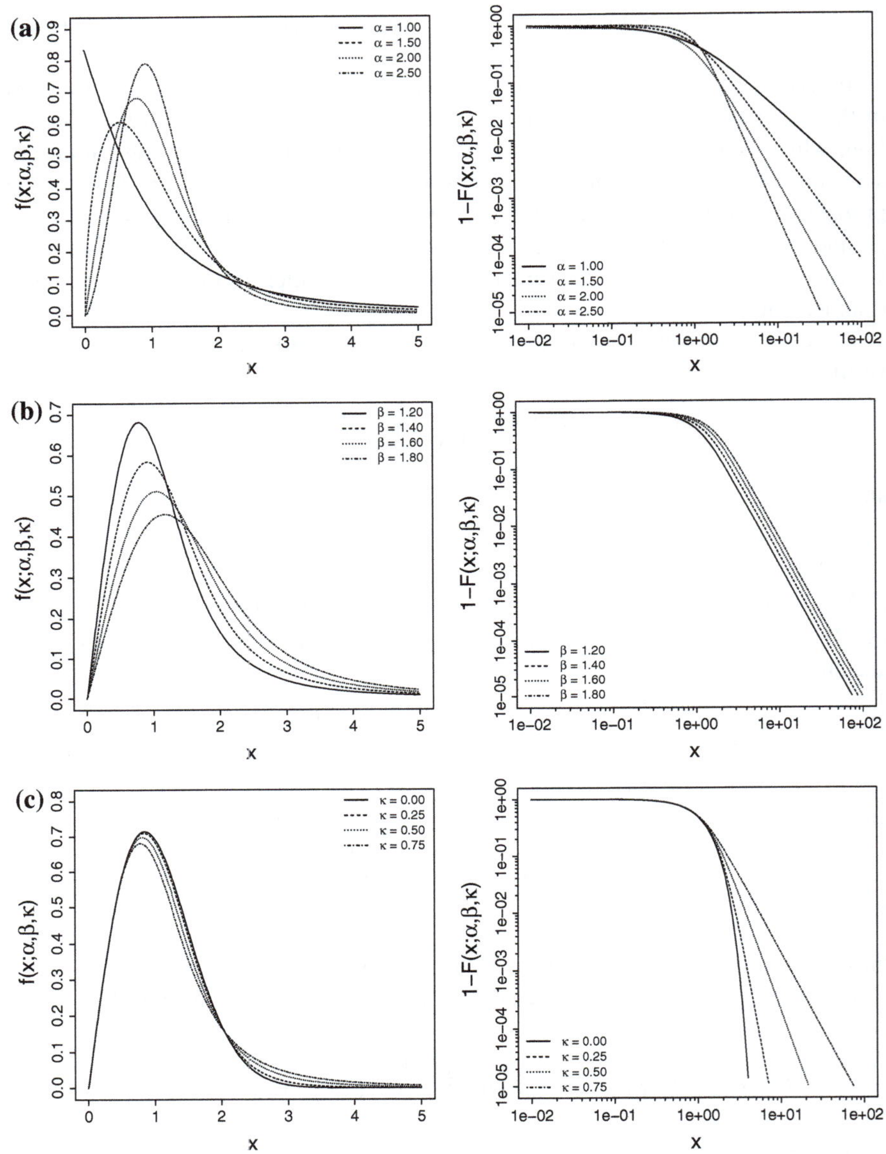

Fig. 3.1 κ-generalized densities (*left*) and log-log complementary distributions (*right*) for some different values of the parameters: **a** $\beta = 1.20$, $\kappa = 0.75$ and $\alpha = 1.00, 1.50, 2.00, 2.50$; **b** $\alpha = 2.00$, $\kappa = 0.75$ and $\beta = 1.20, 1.40, 1.60, 1.80$; **c** $\alpha = 2.00$, $\beta = 1.20$ and $\kappa = 0.00, 0.25, 0.50, 0.75$

an attractive feature for the derivation of Lorenz-ordering results and simulation purposes. From (3.21) we easily determine that the median of the distribution is

$$x_{\text{med}} = \beta \left[\ln_\kappa (2)\right]^{\frac{1}{\alpha}} . \tag{3.22}$$

The mode occurs at

$$x_{\text{mode}} = \beta \left[\frac{\alpha^2 + 2\kappa^2 (\alpha - 1)}{2\kappa^2 \left(\alpha^2 - \kappa^2\right)} \right]^{\frac{1}{2\alpha}} \left\{ \sqrt{1 + \frac{4\kappa^2 \left(\alpha^2 - \kappa^2\right)(\alpha - 1)^2}{\left[\alpha^2 + 2\kappa^2 (\alpha - 1)\right]^2}} - 1 \right\}^{\frac{1}{2\alpha}} \tag{3.23}$$

if $\alpha > 1$; otherwise, the distribution is zero-modal with a pole at the origin.

3.1.3 *Moments and Related Statistics*

The rth-order moment about the origin of the κ-generalized distribution equals

$$\mu_r^{'} = \int_0^\infty x^r f(x; \alpha, \beta, \kappa)\, \mathrm{d}x = \beta^r (2\kappa)^{-\frac{r}{\alpha}} \frac{\Gamma\left(1 + \frac{r}{\alpha}\right)}{1 + \frac{r}{\alpha}\kappa} \frac{\Gamma\left(\frac{1}{2\kappa} - \frac{r}{2\alpha}\right)}{\Gamma\left(\frac{1}{2\kappa} + \frac{r}{2\alpha}\right)}, \tag{3.24}$$

where $\Gamma(\cdot)$ denotes the gamma function, and exists for $-\alpha < r < \frac{\alpha}{\kappa}$. Specifically, $\mu_1^{'} = m$ is the mean of the distribution and

$$\begin{aligned} \sigma^2 &= \mu_2^{'} - m^2 \\ &= \left(\frac{\beta^\alpha}{2\kappa}\right)^{\frac{2}{\alpha}} \left\{ \frac{\Gamma\left(1 + \frac{2}{\alpha}\right)}{1 + 2\frac{\kappa}{\alpha}} \frac{\Gamma\left(\frac{1}{2\kappa} - \frac{1}{\alpha}\right)}{\Gamma\left(\frac{1}{2\kappa} + \frac{1}{\alpha}\right)} - \left[\frac{\Gamma\left(1 + \frac{1}{\alpha}\right)}{1 + \frac{\kappa}{\alpha}} \frac{\Gamma\left(\frac{1}{2\kappa} - \frac{1}{2\alpha}\right)}{\Gamma\left(\frac{1}{2\kappa} + \frac{1}{2\alpha}\right)} \right]^2 \right\} \end{aligned} \tag{3.25}$$

is the variance. Hence, the coefficient of variation is given by

$$CV = \frac{\sigma}{m} = \sqrt{2 \frac{(\alpha + \kappa)^2}{\alpha + 2\kappa} \frac{\Gamma\left(\frac{2}{\alpha}\right)}{\Gamma^2\left(\frac{1}{\alpha}\right)} \frac{\Gamma\left(\frac{1}{2\kappa} - \frac{1}{\alpha}\right)}{\Gamma\left(\frac{1}{2\kappa} + \frac{1}{\alpha}\right)} \frac{\Gamma^2\left(\frac{1}{2\kappa} + \frac{1}{2\alpha}\right)}{\Gamma^2\left(\frac{1}{2\kappa} - \frac{1}{2\alpha}\right)} - 1}. \tag{3.26}$$

It is also possible to define the standardized measures of skewness and kurtosis, respectively given by

$$\gamma_1 = \frac{\mu_3}{\sigma^3} = \frac{\mu_3^{'} - 3\mu_2^{'} m + 2m^3}{\sigma^3} \tag{3.27}$$

and

$$\gamma_2 = \frac{\mu_4}{\sigma^4} = \frac{\mu_4^{'} - 4\mu_3^{'} m - 6\mu_2^{'} m^2 - 3m^4}{\sigma^4}, \tag{3.28}$$

where $\mu_r = \sum_{j=0}^{r} \binom{r}{j} (-1)^{r-j} \mu_j^{'} m^{r-j}$ is the moment about the mean.

3.1.4 Lorenz Curve and Inequality Measures

Since the quantile function of the κ-generalized distribution is available in closed form, its normalized integral, the Lorenz (1905) curve

$$L(u) = \frac{1}{m}\int_0^u F^{-1}(t)\,\mathrm{d}t, \quad u \in [0,1], \tag{3.29}$$

can be expressed analytically (Gastwirth 1971). Therefore, we have

$$\begin{aligned} L(u) = 1 - \frac{1+\frac{\kappa}{\alpha}}{2\Gamma\left(\frac{1}{\alpha}\right)} \frac{\Gamma\left(\frac{1}{2\kappa}+\frac{1}{2\alpha}\right)}{\Gamma\left(\frac{1}{2\kappa}-\frac{1}{2\alpha}\right)} \left\{ \frac{2\alpha(1-u)}{(2\kappa)^{-\frac{1}{\alpha}}} \left[\ln_\kappa\left(\frac{1}{1-u}\right)\right]^{\frac{1}{\alpha}} \right. \\ \left. + B_x\left(\frac{1}{2\kappa}-\frac{1}{2\alpha}, \frac{1}{\alpha}\right) + B_x\left(\frac{1}{2\kappa}-\frac{1}{2\alpha}+1, \frac{1}{\alpha}\right)\right\}, \end{aligned} \tag{3.30}$$

where $B_x(\cdot,\cdot)$ is the incomplete beta function with $x = (1-u)^{2\kappa}$. The equivalent expression

$$L(u) = I_x\left(1+\frac{1}{\alpha}, \frac{1}{2\kappa}-\frac{1}{2\alpha}\right), \quad x = 1-(1-u)^{2\kappa}, \tag{3.31}$$

has been independently derived by Okamoto (2013), where $I_x(\cdot,\cdot)$ is the regularized incomplete beta function defined in terms of the incomplete beta function and the complete beta function—that is, $I_x(\cdot,\cdot) = \frac{B_x(\cdot,\cdot)}{B(\cdot,\cdot)}$. Clearly, the curve exists if and only if $\frac{\alpha}{\kappa} > 1$. The use of (3.30) and (3.31) can be done analytically.

Given two κ-generalized distributions X_1 and X_2, the following theorem gives the parameter constellations for which their Lorenz curves do not intersect and we have $X_1 \leq_L X_2$, i.e. when the Lorenz curve of X_1 lies nowhere below that of X_2 (in symbols: $L_{X_1}(u) \geq L_{X_2}(u)$, for all $u \in [0,1]$) and consequently X_1 exhibits less inequality than X_2 in the Lorenz sense:

Theorem 3.1 *Let $X_i \sim \kappa\text{-gen}(\alpha_i, \beta_i, \kappa_i)$, $i = 1, 2$, be κ-generalized distributions. Then*

$$X_1 \leq_L X_2 \Longleftrightarrow \alpha_1 \geq \alpha_2 \text{ and } \frac{\alpha_1}{\kappa_1} \geq \frac{\alpha_2}{\kappa_2}. \tag{3.32}$$

Figure 3.2 provides an illustration of (3.32), showing that the less unequal distribution (in the Lorenz sense) always exhibits lighter tails.[11]

---

[11] Within three-parameter families, the Lorenz-ordering has been settled by Taillie (1981) and Wilfling (1996b) for the generalized gamma distribution and by Wilfling and Krämer (1993) and Kleiber (1996), respectively, for the most popular Singh-Maddala and Dagum type I distributions. Both these distributions are special cases of McDonald (1984) four-parameter GB2 distribution. For this family, Wilfling (1996a) was able to derive a necessary and also a sufficient condition.

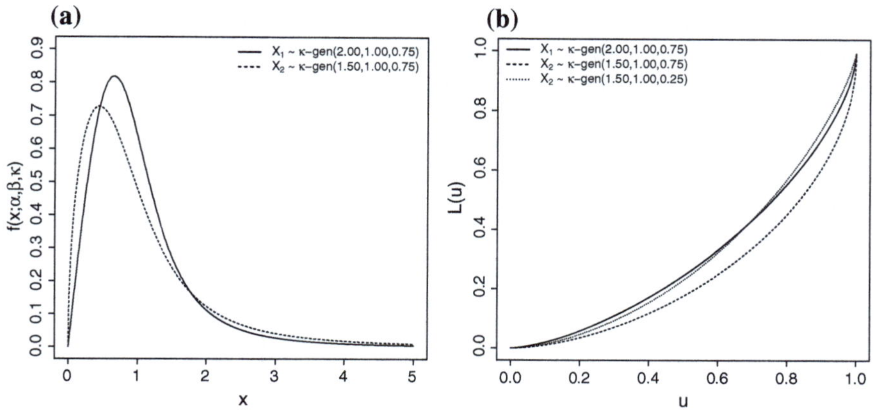

Fig. 3.2 Tails (**a**) and Lorenz curves (**b**) for two κ-generalized distributions. As shown by the *dotted line*, the Lorenz curves intersect if the parameters are not chosen according to Theorem 3.1

The proof of Theorem 3.1 proceeds as follows:

Proof Since the Lorenz curve is invariant under scale changes, the parameter β does not enter expression (3.30) and can be chosen as 1 without loss of generality. Furthermore, we state the following:

Lemma 3.1 (Hardy et al. 1929; see also Marshall and Olkin 1979) *Let X_i, $i = 1, 2$, be positive random variables having finite mean m_{X_i}. Then $X_1 \leq_L X_2$ if and only if*

$$\mathbb{E}\left[\psi\left(\frac{X_1}{m_{X_1}}\right)\right] \leq \mathbb{E}\left[\psi\left(\frac{X_2}{m_{X_2}}\right)\right] \tag{3.33}$$

for all continuous and convex functions $\psi : \mathbb{R}_+ \to \mathbb{R}$ for which the expectation exists.

Now we can prove the first direction of Theorem 3.1, i.e. necessity of conditions $\alpha_1 \geq \alpha_2$ and $\frac{\alpha_1}{\kappa_1} \geq \frac{\alpha_2}{\kappa_2}$ for Lorenz-ordering of κ-generalized distributions. Consider the family of continuous convex functions

$$\Psi(x) = \frac{x^{t+1} - 1}{t(t+1)}, \quad x > 0, \quad -\infty < t < \infty, \quad t \neq -1, 0. \tag{3.34}$$

As can be seen in Taillie (1981), corresponding to Ψ one can obtain the family of Lorenz-ordering preserving inequality measures

$$H_t(X) = \mathbb{E}\left[\Psi\left(\frac{X}{m_X}\right)\right] = \frac{1}{t(t+1)}\left[\frac{\mathbb{E}\left(X^{t+1}\right)}{m_X^{t+1}} - 1\right], \tag{3.35}$$

where $x > 0$, $-\infty < t < \infty$ and $t \neq -1, 0$. This family includes some standard indices as special cases, such as one-half the squared coefficient of variation (H_1) and the Theil index (H_0). From Lemma 3.1 we know that

$$H_t(X_1) \leq H_t(X_2), \tag{3.36}$$

where

$$\begin{aligned} H_t(X_1) &= \frac{1}{t(t+1)} \left\{ \frac{(2\kappa_1)^{-\frac{t+1}{\alpha_1}} \frac{\Gamma\left(1+\frac{t+1}{\alpha_1}\right)}{1+\frac{t+1}{\alpha_1}\kappa_1} \frac{\Gamma\left(\frac{1}{2\kappa_1}-\frac{t+1}{2\alpha_1}\right)}{\Gamma\left(\frac{1}{2\kappa_1}+\frac{t+1}{2\alpha_1}\right)}}{\left[(2\kappa_1)^{-\frac{1}{\alpha_1}} \frac{\Gamma\left(1+\frac{1}{\alpha_1}\right)}{1+\frac{\kappa_1}{\alpha_1}} \frac{\Gamma\left(\frac{1}{2\kappa_1}-\frac{1}{2\alpha_1}\right)}{\Gamma\left(\frac{1}{2\kappa_1}+\frac{1}{2\alpha_1}\right)}\right]^{t+1}} - 1 \right\} \\ &= \frac{1}{t(t+1)} \left\{ \frac{\Gamma\left(1+\frac{t+1}{\alpha_1}\right)}{1+\frac{t+1}{\alpha_1}\kappa_1} \frac{\Gamma\left(\frac{1}{2\kappa_1}-\frac{t+1}{2\alpha_1}\right)}{\Gamma\left(\frac{1}{2\kappa_1}+\frac{t+1}{2\alpha_1}\right)} \left[\frac{1+\frac{\kappa_1}{\alpha_1}}{\Gamma\left(1+\frac{1}{\alpha_1}\right)} \frac{\Gamma\left(\frac{1}{2\kappa_1}+\frac{1}{2\alpha_1}\right)}{\Gamma\left(\frac{1}{2\kappa_1}-\frac{1}{2\alpha_1}\right)} \right]^{t+1} \right. \\ &\quad \left. -1 \right\} \end{aligned} \tag{3.37}$$

follows from (3.24) and is not defined outside the interval $\left(-\alpha_1 - 1, \frac{\alpha_1}{\kappa_1} - 1\right)$. A similar expression obtains for $H_t(X_2)$. Recalling that $\lim_{z\to 0} \Gamma(z) = +\infty$, the terms $\Gamma\left(1 + \frac{t+1}{\alpha_1}\right)$ and $\Gamma\left(\frac{1}{2\kappa_1} - \frac{t+1}{2\alpha_1}\right)$ tend to infinity as t approaches $-\alpha_1 - 1$ from above and $\frac{\alpha_1}{\kappa_1} - 1$ from below, respectively. Similarly, $H_t(X_2)$ is not defined outside the interval $\left(-\alpha_2 - 1, \frac{\alpha_2}{\kappa_2} - 1\right)$, and tends to infinity as $t \to -\alpha_2 - 1$ and $t \to \frac{\alpha_2}{\kappa_2} - 1$. Hence, in conjunction with (3.36), it follows that $-\alpha_1 - 1 \leq -\alpha_2 - 1$ and $\frac{\alpha_1}{\kappa_1} - 1 \geq \frac{\alpha_2}{\kappa_2} - 1$, or equivalently $\alpha_1 \geq \alpha_2$ and $\frac{\alpha_1}{\kappa_1} \geq \frac{\alpha_2}{\kappa_2}$.

It remains to prove the reverse direction, i.e. that the conditions of Theorem 3.1 are also sufficient. Since the quantile function is available in a simple closed form, we can check for star-shaped ordering $X_1 \leq_* X_2$ (e.g. Arnold 1987) by verifying that the derivative with respect to u of the ratio

$$\frac{F_{X_2}^{-1}(u; \alpha_2, \kappa_2)}{F_{X_1}^{-1}(u; \alpha_1, \kappa_1)} = \frac{\left[\ln_{\kappa_2}\left(\frac{1}{1-u}\right)\right]^{\frac{1}{\alpha_2}}}{\left[\ln_{\kappa_1}\left(\frac{1}{1-u}\right)\right]^{\frac{1}{\alpha_1}}} = \frac{\left[\frac{(1-u)^{-\kappa_2}-(1-u)^{\kappa_2}}{2\kappa_2}\right]^{\frac{1}{\alpha_2}}}{\left[\frac{(1-u)^{-\kappa_1}-(1-u)^{\kappa_1}}{2\kappa_1}\right]^{\frac{1}{\alpha_1}}} \tag{3.38}$$

is non-negative. After some straightforward manipulations, one ends up with the result

$$\frac{\alpha_1}{\kappa_1}\left[\frac{(1-u)^{-\kappa_1} - (1-u)^{\kappa_1}}{(1-u)^{-\kappa_1} + (1-u)^{\kappa_1}}\right] \geq \frac{\alpha_2}{\kappa_2}\left[\frac{(1-u)^{-\kappa_2} - (1-u)^{\kappa_2}}{(1-u)^{-\kappa_2} + (1-u)^{\kappa_2}}\right], \tag{3.39}$$

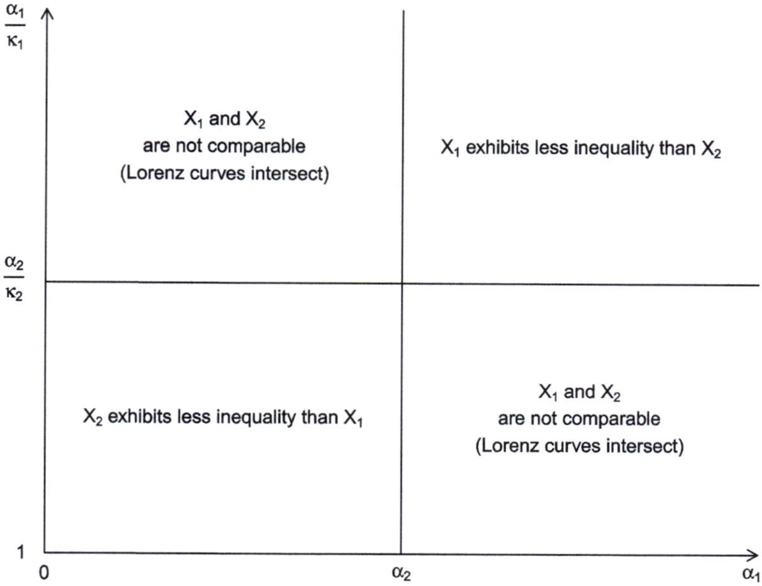

Fig. 3.3 Lorenz-ordering within the κ-generalized family

which holds true assuming that $\alpha_1 \geq \alpha_2$ and $\frac{\alpha_1}{\kappa_1} \geq \frac{\alpha_2}{\kappa_2}$. Therefore, conditions (3.32) are sufficient for the star-shaped order of X_1 with respect to X_2, which in turn implies the Lorenz-ordering $X_1 \leq_L X_2$ (e.g. Chandra and Singpurwalla 1981).

Figure 3.3 shows the decomposition of $(\alpha_1, \frac{\alpha_1}{\kappa_1})$-space into regions where X_1 dominates or is dominated by X_2, given α_2 and $\frac{\alpha_2}{\kappa_2}$, and where instead the two distributions are not comparable because their Lorenz curves cross. □

As regards scalar measures of inequality, the well-known Gini (1914) coefficient can be derived using the representation in terms of order statistics due to Arnold and Laguna (1977), i.e. $G = 1 - m^{-1} \int_0^\infty [1 - F(x)]^2 \, dx$; this yields[12]

$$G = 1 - \frac{2\alpha + 2\kappa}{2\alpha + \kappa} \frac{\Gamma\left(\frac{1}{\kappa} - \frac{1}{2\alpha}\right) \Gamma\left(\frac{1}{2\kappa} + \frac{1}{2\alpha}\right)}{\Gamma\left(\frac{1}{\kappa} + \frac{1}{2\alpha}\right) \Gamma\left(\frac{1}{2\kappa} - \frac{1}{2\alpha}\right)}. \tag{3.40}$$

Although commonly used, the Gini is but one of many measures of inequality that are available, and it incorporates particular assumptions about the way in which

[12]Using Stirling approximation for the gamma function, $\Gamma(z) \approx \sqrt{2\pi} z^{z-\frac{1}{2}} \exp(-z)$, and taking the limit as $\kappa \to 0$ in Eq. (3.40), one arrives after some simplification at $G = 1 - 2^{-\frac{1}{\alpha}}$, which is the explicit form of the Gini coefficient for the Weibull distribution (e.g. Kleiber and Kotz 2003, p. 177). Since the exponential distribution is a special case of the Weibull with shape parameter equal to 1, one directly determines that for $\kappa \to 0$ and $\alpha = 1$ the exponential law is also a special limiting case of the κ-generalized distribution, with a "true" value of the Gini coefficient equal to one half (e.g. Drăgulescu and Yakovenko 2001).

income differences in different parts of the distribution are summarized—it tends to be most sensitive to transfers around the middle of the income distribution and least sensitive to transfers among the very rich or the very poor (Allison 1978). In other forms of income distribution research, the generalized entropy class of inequality measures (Cowell 1980a, b; Cowell and Kuga 1981a, b; Shorrocks 1980) provides a full range of bottom- to top-sensitive indices that are widely used by analysts to assess inequality in different parts of the income distribution. In terms of the κ-generalized parameters, this class of inequality indices is expressed as

$$GE(\theta) = \frac{1}{\theta^2 - \theta}\left\{\left(\frac{\beta}{m}\right)^{\theta}\left[\frac{(2\kappa)^{-\frac{\theta}{\alpha}}}{1+\frac{\theta}{\alpha}\kappa}\frac{\Gamma\left(\frac{1}{2\kappa}-\frac{\theta}{2\alpha}\right)}{\Gamma\left(\frac{1}{2\kappa}+\frac{\theta}{2\alpha}\right)}\Gamma\left(1+\frac{\theta}{\alpha}\right)\right]-1\right\}, \tag{3.41}$$

where $\theta = 0, 1$. Equation (3.41) defines a class because the index $GE(\theta)$ assumes different forms depending on the value assigned to θ, the parameter that characterizes the sensitivity of $GE(\theta)$ to income differences in different parts of the income distribution—the more positive that θ is, the more sensitive is $GE(\theta)$ to income differences at the top of the distribution; the more negative that θ is, the more sensitive is $GE(\theta)$ to income differences at the bottom of the distribution. In applied work, two limiting cases of (3.41)—the more middle-sensitive indices—are of particular interest for inequality measurement: the mean logarithmic deviation index

$$MLD = \lim_{\theta\to 0} GE(\theta) = \frac{1}{\alpha}\left[\gamma + \psi\left(\frac{1}{2\kappa}\right) + \ln(2\kappa) - \alpha\ln\left(\frac{\beta}{m}\right) + \kappa\right], \tag{3.42}$$

where $\gamma = -\psi(1)$ is the Euler-Mascheroni constant and $\psi(z) = \Gamma'(z)/\Gamma(z)$ is the digamma function, and the Theil (1967) index

$$\begin{aligned} T &= \lim_{\theta\to 1} GE(\theta) \\ &= \frac{1}{\alpha}\left[\psi\left(1+\frac{1}{\alpha}\right) - \frac{1}{2}\psi\left(\frac{1}{2\kappa}-\frac{1}{2\alpha}\right) - \frac{1}{2}\psi\left(\frac{1}{2\kappa}+\frac{1}{2\alpha}\right)\right. \\ &\quad \left. - \ln(2\kappa) + \alpha\ln\left(\frac{\beta}{m}\right) - \frac{\alpha\kappa}{\alpha+\kappa}\right]. \end{aligned} \tag{3.43}$$

Expression for each index other than for the cases (3.42) and (3.43) can be derived by straightforward substitution. In particular, the bottom-sensitive index is given by

$$GE(-1) = -\frac{1}{2} + \frac{\Gamma\left(1+\frac{1}{\alpha}\right)\Gamma\left(1-\frac{1}{\alpha}\right)}{2\left[1-\left(\frac{\kappa}{\alpha}\right)^2\right]}, \tag{3.44}$$

whereas the expression for the top-sensitive index is

$$GE(2) = \frac{1}{2}\left\{2\frac{(\alpha+\kappa)^2}{\alpha+2\kappa}\frac{\Gamma\left(\frac{2}{\alpha}\right)}{\Gamma^2\left(\frac{1}{\alpha}\right)}\frac{\Gamma\left(\frac{1}{2\kappa}-\frac{1}{\alpha}\right)}{\Gamma\left(\frac{1}{2\kappa}+\frac{1}{\alpha}\right)}\frac{\Gamma^2\left(\frac{1}{2\kappa}+\frac{1}{2\alpha}\right)}{\Gamma^2\left(\frac{1}{2\kappa}-\frac{1}{2\alpha}\right)}-1\right\} \qquad (3.45)$$
$$= \frac{1}{2}CV^2.$$

Finally, the Atkinson (1970) class of inequality indices can be easily computed from (3.41) by exploiting the relationship (Cowell 2011; Jenkins 2009)

$$A(\epsilon) = 1 - \left[\epsilon(\epsilon-1)GE(1-\epsilon)+1\right]^{\frac{1}{1-\epsilon}}, \qquad (3.46)$$

where $\epsilon = 1-\theta$, $\epsilon > 0$ and $\epsilon \neq 1$, is the inequality aversion parameter. As ϵ rises, $A(\epsilon)$ becomes more sensitive to transfers among lower incomes and less sensitive to transfers among top income recipients (Allison 1978). The limiting form of the equation above is $A(1) = 1 - \exp(-MLD)$.[13]

3.1.5 Economic Interpretation of the Parameters

Dagum (1977) proposed a large set of ideal properties to motivate and guide the identification of a functional form as a model of personal income distribution. This set includes, among others, the requirement that all parameters of the specified model have a well-defined economic meaning.

Two general categories of parameters encountered often are the scale parameter and the inequality parameters. The former is related to the income measurement unit, while the latter are dimensionless parameters related to the inequality of the income distribution—as they enter into several inequality measures that can be derived from a given income distribution function.

As already noted in Sect. 3.1.2, the parameter β of the κ-generalized model is a characteristic scale having the same dimension of income. For this reason, it accounts for the monetary scale, and can be used as a tool to adjust for inflation and to facilitate cross-country comparisons of income distributions that are expressed in different monetary units. Increments (reductions) in the monetary unit generate a global increase (decrease) of each and every one of the incomes and, therefore, of the average income. In this way, inequality measures remain invariant.

[13]Notice that all the measures considered here are functions of the distribution moments, whose existence depends on some conditions guaranteeing the convergence of relevant integrals. As a matter of example, the Gini coefficient (3.40) exists provided the mean of the distribution $m = \int_0^\infty xf(x;\alpha,\beta,\kappa)\,dx$ converges, which is true when $\frac{\alpha}{\kappa} > 1$. As shown by Kleiber (1997), the problem of existence of popular inequality measures is common to various parametric models of income distribution.

By contrast, α and κ are scale-free parameters: changes in the monetary unit yield β modifications but leave α and κ invariant. The α and κ modifications affect instead the shape of the distribution: as shown in Fig. 3.1, the region around the origin of the κ-generalized distribution is governed by α, the upper tail by both α and κ.[14] In particular, increasing κ leads to a thicker upper tail, whereas increasing α lowers both the tails and yields a greater concentration of probability mass around the peak of the distribution.

As elaborated on in the previous section, the dimensionless parameters α and κ also enter the explicit mathematical solution for the Gini index and any of the other inequality measures that can be derived from the κ-generalized model of income distribution. However, the influence of each parameter on these measures is different. For instance, Clementi et al. (2010) obtained a first interpretation of κ as an "inequality" parameter and α as an "equality" parameter after verifying that the Gini coefficient (3.40) increases with the former and decreases with the latter. In order to corroborate this interpretation, we carry out a graphic analysis similar to that performed by Jäntti and Jenkins (2010) and Pérez and Alaiz (2011) using the parameters of the Singh-Maddala and Dagum models, respectively.

Panel (a) of Fig. 3.4 shows different combinations of the α and κ parameters which give a specific value of the Gini coefficient.[15] From this figure, the positive (negative) relation between the Gini coefficient and the inequality (equality) parameter κ (α) is confirmed: the contours nearest the top left-hand (bottom right-hand) corner correspond to the highest (lowest) Gini values and the largest values of the parameter, while the opposite happens with the contours located at the bottom right-hand (top left-hand) corner. Furthermore, higher κ values holding α constant are associated with increased inequality, whereas higher values of α holding κ constant lead to declines in inequality.

A similar pattern holds for the other inequality measures in panels (b) to (f) of the same figure: as suggested indeed by the contours in the parameter plane, increases in α and decreases in κ are associated with declining levels of inequality, whereas the profile of inequality is reversed if one moves along the opposite direction of change.

To deepen the interpretation of the parameters α and κ, we also analyze how these affect specific parts of the income distribution, which cannot be assessed when using global inequality indicators such as, for instance, the Gini index. In particular, we analyze the specific influence of each parameter in the lower and upper tails of the distribution, in order to explain changes in income distribution when α and κ move. For this purpose, we use the elasticity function of percentiles, which allows us to

[14] A similar observation was reported by Kleiber (1996) in his study on the relationship between the Dagum and Singh-Maddala models. He notes, in fact, that the upper tail of the Singh-Maddala model is governed by two parameters and the region near the origin (where there are more individuals or families) is controlled only by one. The opposite happens, instead, with the Dagum model.

[15] The values of the parameters have been chosen similar to those estimated by Clementi et al. (2010, Tables 2–4, pp. 571–576) using Italian household income data for the years 1989 to 2006.

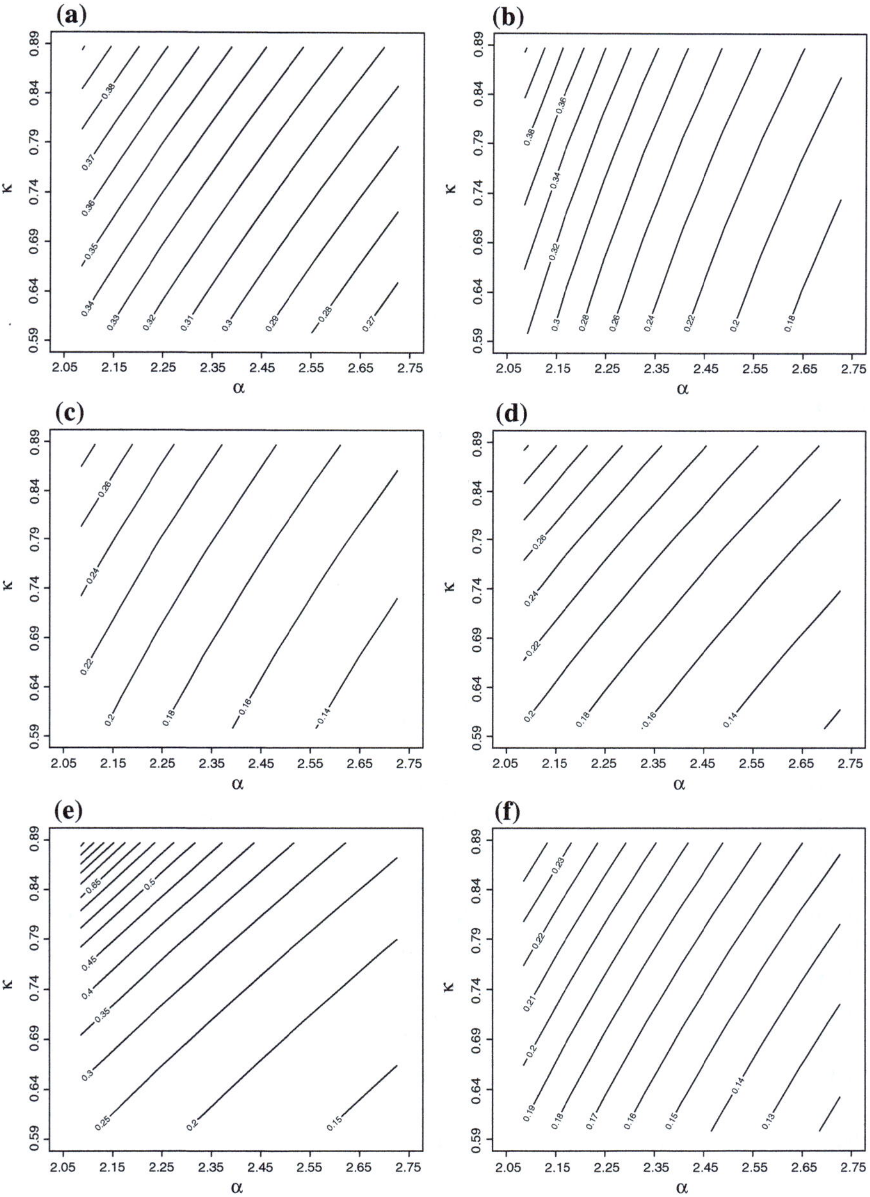

Fig. 3.4 Sensibility of κ-generalized inequality measures to α and κ parameters: **a** Gini; **b** bottom-sensitive *GE* index; **c** mean logarithmic deviation; **d** Theil; **e** top-sensitive *GE* index. **f** Atkinson (using the inequality aversion parameter $\epsilon = 1$)

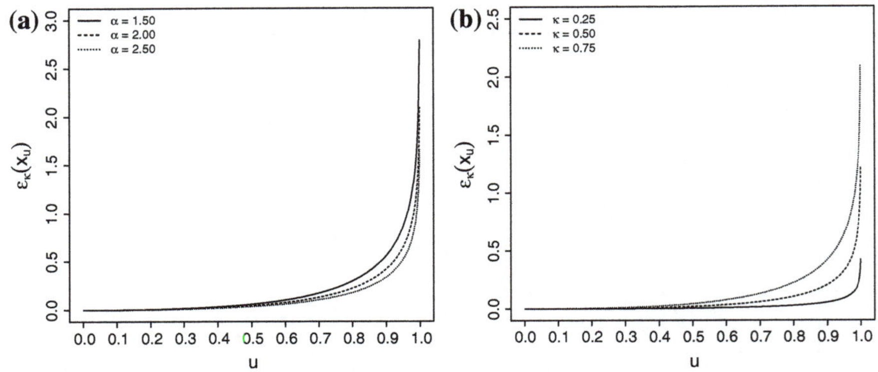

Fig. 3.5 Elasticity function of percentiles with respect to κ parameter for different combinations of α and κ: **a** $\kappa = 0.75$ and $\alpha = 1.50, 2.00, 2.50$; **b** $\alpha = 2.00$ and $\kappa = 0.25, 0.50, 0.75$

study the influence of parameter variations on the distribution for different percentile values.[16]

The elasticity function of the uth percentile with respect to the κ parameter is given by the following expression

$$\epsilon_\kappa(x_u) = \frac{\partial x_u}{\partial \kappa}\frac{\kappa}{x_u} = \frac{1}{\alpha}\left\{\kappa \ln\left(\frac{1}{1-u}\right)\left[\frac{\left(\frac{1}{1-u}\right)^{2\kappa}+1}{\left(\frac{1}{1-u}\right)^{2\kappa}-1}\right] - 1\right\} \tag{3.47}$$

and is found to be positive for every u irrespective of the values the parameters α and κ assume (Fig. 3.5). In particular, it is shown that $\epsilon_\kappa(x_u)$ is an increasing function in u, and so the increments of upper percentiles will be stronger with increasing values of u. Panel (b) of Fig. 3.5 also confirms that the κ parameter controls the upper tail of the distribution in such a way that the larger the value of the parameter is, the higher the upper percentile increments are. This means that if κ increases, the thickness of the upper tail of the income distribution would also increase, leading any of the inequality measures considered before to raise.

As we have mentioned, Clementi et al. (2010), in a preliminary analysis, conclude that α is an equality parameter because the Gini coefficient is a decreasing function of it. In order to complete this interpretation, we analyze the elasticity function of the percentiles with respect to α, defined as

$$\epsilon_\alpha(x_u) = \frac{\partial x_u}{\partial \alpha}\frac{\alpha}{x_u} = -\frac{1}{\alpha}\ln\left[\ln_\kappa\left(\frac{1}{1-u}\right)\right]. \tag{3.48}$$

[16]For similar analyses using the elasticity function of percentiles with respect to the different parameters of the Dagum model see García et al. (2006) and Pérez and Alaiz (2011).

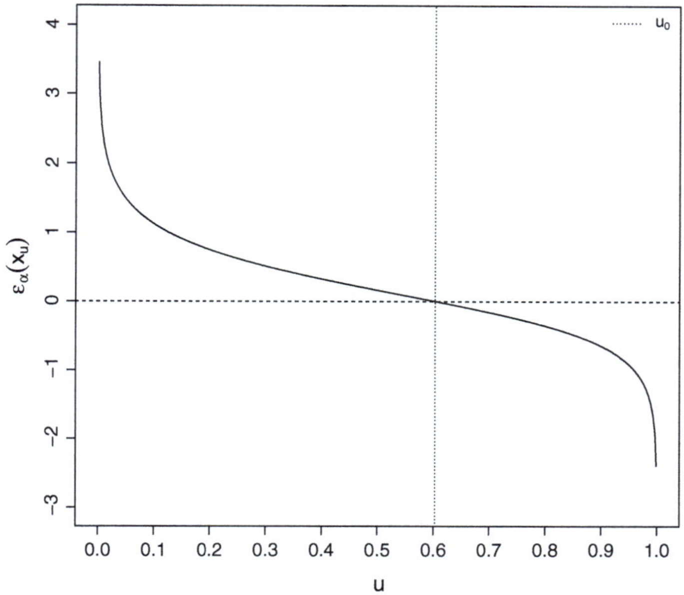

Fig. 3.6 Elasticity function of percentiles with respect to α parameter. The *solid line* has been drawn for $\alpha = 2.00$ and $\kappa = 0.75$

From this expression it can be deduced that

$$\epsilon_\alpha (x_u) > 0 \quad \text{if} \quad u < 1 - \left(\kappa + \sqrt{\kappa^2 + 1}\right)^{-\frac{1}{\kappa}}, \tag{3.49}$$

and

$$\epsilon_\alpha (x_u) < 0 \quad \text{if} \quad u > 1 - \left(\kappa + \sqrt{\kappa^2 + 1}\right)^{-\frac{1}{\kappa}}. \tag{3.50}$$

This means that any increase in α involves improvements of the income percentiles below $u_0 = 1 - \left(\kappa + \sqrt{\kappa^2 + 1}\right)^{-\frac{1}{\kappa}}$ and reductions of income percentiles above u_0. As depicted in Fig. 3.6, $\epsilon_\alpha (x_u)$ is a decreasing function in u, and so the reductions (increments) of upper (lower) percentiles will be stronger (weaker) with increasing values of u. Besides, the graph confirms that the α parameter affects both tails of the distribution, other than the middle range of income. We can thus conclude that α is an equality parameter because it causes upper percentile reductions and lower-middle percentile increments. Holding κ constant, this effect on the income distribution also causes a decrease in income inequality.

The double effect of the α parameter on the upper and lower percentiles can also provoke a less bi-polarized income distribution, with more individuals or families near middle-income levels. Given that polarization is related to the concentration of the distribution on several poles, whereas inequality relates to the overall dispersion, it is interesting to complement the study of the α influence with an analysis of its effect on a polarization measure. To this end, we use the Wolfson (1994) polarization index, which is a bi-polarization measure. It can be expressed as a function of the Esteban and Ray (1994) polarization index when only two groups (respectively below and above the median) are considered.[17] The expression of the Wolfson measure is the following:

$$W = 2\left[1 - 2L\left(0.5\right) - G\right]\frac{m}{x_{\mathrm{med}}}, \tag{3.51}$$

where $L\left(0.5\right)$ represents the income share of the population whose income is lower than the median (the poor half of the population), G is the Gini coefficient, and m and x_{med} are the mean and the median of the distribution. The index (3.51) ranges within the interval [0, 1], being equal to 0 in case of a perfectly equal distribution—all the incomes are equal—and equal to 1 for a perfect bimodal distribution, where half of the population has no income and each member of the other half has income equal to twice the mean income.

We can express this index as a function of the κ-generalized parameters in the following way

$$\begin{aligned} W = {} & 2\left(2^{\kappa} - 2^{-\kappa}\right)^{-\frac{1}{\alpha}} \frac{\Gamma\left(1 + \frac{1}{\alpha}\right)}{1 + \frac{\kappa}{\alpha}} \frac{\Gamma\left(\frac{1}{2\kappa} - \frac{1}{2\alpha}\right)}{\Gamma\left(\frac{1}{2\kappa} + \frac{1}{2\alpha}\right)} \left[\frac{2\alpha + 2\kappa}{2\alpha + \kappa} \frac{\Gamma\left(\frac{1}{\kappa} - \frac{1}{2\alpha}\right)}{\Gamma\left(\frac{1}{\kappa} + \frac{1}{2\alpha}\right)}\right. \\ & \left. \cdot \frac{\Gamma\left(\frac{1}{2\kappa} + \frac{1}{2\alpha}\right)}{\Gamma\left(\frac{1}{2\kappa} - \frac{1}{2\alpha}\right)} - 2I_x\left(1 + \frac{1}{\alpha}, \frac{1}{2\kappa} - \frac{1}{2\alpha}\right)\right], \end{aligned} \tag{3.52}$$

where $x = 1 - 0.5^{2\kappa}$. We also show, in Fig. 3.7, different combinations of the α and κ parameters which give specific values of the measure (3.52). We can see that an increase in α, holding κ constant, yields decreases in income polarization as this parameter provokes both increases in the lower incomes as well as decreases in the upper ones. On the other hand, the parameter κ considerably influences incomes in the

[17]The Wolfson index of bi-polarization has been also reformulated in terms of the between-group and within-group Gini coefficients (Rodríguez and Salas 2003). More precisely, there are two characteristics that are considered as being intrinsic to the notion of bi-polarization. The first one, "increased spread", implies that moving from the central value (median) to the extreme points of the income distribution makes the distribution more polarized than before. In other words, increments (reductions) in incomes above (below) the median will widen the distribution, that is extend the distance between the groups below and above the median and hence increase the degree of bi-polarization. On the other hand, "increased bi-polarity" refers to the case where incomes on the same side of the median get closer to each other. Since the distance between the incomes below or above the median has been reduced, this is assumed to increase bi-polarization. Thus, bi-polarization involves both an inequality-like component, the "increased spread" principle, which increases both inequality and polarization, and an equality-like component, the "increased bi-polarity" criterion, which increases polarization but lowers any inequality measure that fulfills the Pigou-Dalton transfer principle—the requirement under which inequality decreases when a transfer is made from a richer to a poorer individual without reversing their pairwise ranking. This shows that although there is complementarity between polarization and inequality, there are differences as well. See for instance Deutsch et al. (2013) for a thorough discussion.

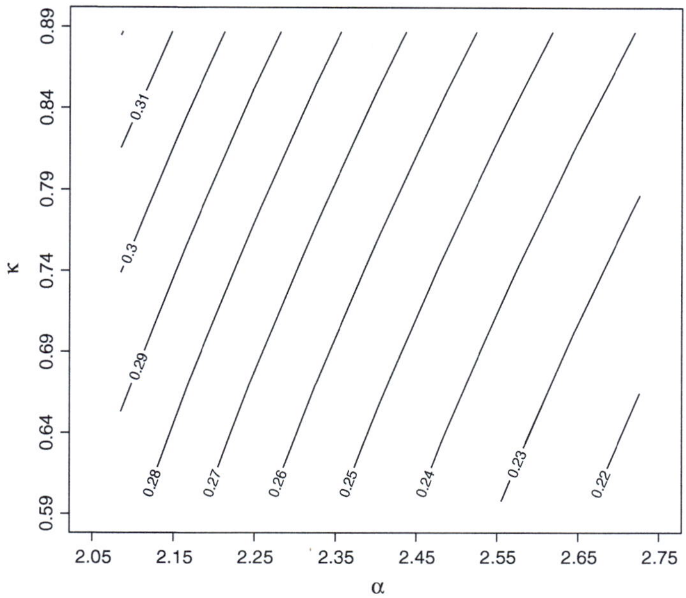

Fig. 3.7 Sensibility of Wolfson bi-polarization measure to α and κ parameters. The values of the parameters have been chosen similar to those estimated by Clementi et al. (2010, Tables 2–4, pp. 571–576) using Italian household income data for the years 1989–2006

upper tail of the distribution. Hence, if this parameter increases, all upper percentiles will increase and this effect will cause an increment in income polarization. Thus, as in the case of the Gini coefficient and the other income inequality measures, an increase in α and/or a decrease in κ will lower income polarization, while the opposite happens when α decreases and/or κ increases.

3.1.6 Estimation

Parameter estimation for the κ-generalized distribution can be performed using the maximum likelihood method, which yields estimators with good statistical properties (Rao 1973; Ghosh 1994). Assuming that all sample observations $\mathbf{x} = \{x_1, \ldots, x_n\}$ are independent, the likelihood function is

$$L(\mathbf{x}; \boldsymbol{\theta}) = \prod_{i=1}^{n} f(x_i; \boldsymbol{\theta})^{w_i} = \prod_{i=1}^{n} \left\{ \frac{\alpha}{\beta} \left(\frac{x_i}{\beta} \right)^{\alpha-1} \frac{\exp_\kappa \left[-(x_i/\beta)^\alpha \right]}{\sqrt{1+\kappa^2 (x_i/\beta)^{2\alpha}}} \right\}^{w_i}, \tag{3.53}$$

where $f(x_i; \boldsymbol{\theta})$ is the probability distribution function, $\boldsymbol{\theta} = \{\alpha, \beta, \kappa\}$ is the unknown parameters vector, w_i is the weight of ith observation and n is the sample size. This leads to the problem of solving the partial derivatives of the log-likelihood function

$$l(\mathbf{x};\boldsymbol{\theta}) = \ln\left[L(\mathbf{x};\boldsymbol{\theta})\right] = \sum_{i=1}^{n} w_i \ln\left[f(x_i;\boldsymbol{\theta})\right] \tag{3.54}$$

with respect to α, β and κ, which corresponds to finding the solution of the following non-linear system of equations

$$\sum_{i=1}^{n} w_i \frac{\partial}{\partial\alpha} \ln\left[f(x_i;\boldsymbol{\theta})\right] = 0, \tag{3.55}$$

$$\sum_{i=1}^{n} w_i \frac{\partial}{\partial\beta} \ln\left[f(x_i;\boldsymbol{\theta})\right] = 0, \tag{3.56}$$

$$\sum_{i=1}^{n} w_i \frac{\partial}{\partial\kappa} \ln\left[f(x_i;\boldsymbol{\theta})\right] = 0. \tag{3.57}$$

However, obtaining explicit expressions for the maximum likelihood estimators of the three parameters by resolution of the above equations is difficult, making direct analytical solutions intractable. The derivation of any of these estimators, therefore, generally involves the use of numerical optimization algorithms.[18]

3.2 Empirical Results and Comparisons to Alternative Income Distributions

In what follows, we focus on the ability of the proposed model to describe real-world income distributions. First, a brief description of the income data used for parameter estimation is provided. Then we consider a two-country case study for Italy and the United States as an empirical illustration of the performance of the κ-generalized distribution. To conclude, the relative merits of alternative models for the income size distribution are compared by fitting them to all the most recent country data sets available to date in the income micro-database being used.

[18]Clementi et al. (2007, 2008, 2009) use a slightly different parametrization of Eq. (3.3), namely $f(x;\alpha,\lambda,\kappa) = \alpha\lambda x^{\alpha-1}\frac{\exp_\kappa(-\lambda x^\alpha)}{\sqrt{1+\kappa^2\lambda^2x^{2\alpha}}}$, where $\lambda = \beta^{-\alpha}$ and x is defined as $x = \frac{z}{\langle z\rangle}$, being z the absolute personal income and $\langle z\rangle$ its mean value. Taking into account the meaning of the variable x, the mean value results to be equal to unity, i.e. $m = \int_0^\infty xf(x;\alpha,\lambda,\kappa)\,\mathrm{d}x = 1$. The latter relationship permits to express the scale parameter λ as a function of the shape parameters α and κ, obtaining $\lambda = \frac{1}{2\kappa}\left[\frac{\Gamma\left(\frac{1}{\alpha}\right)}{\kappa+\alpha}\frac{\Gamma\left(\frac{1}{2\kappa}-\frac{1}{2\alpha}\right)}{\Gamma\left(\frac{1}{2\kappa}+\frac{1}{2\alpha}\right)}\right]^\alpha$. In this way, the problem to determine the values of the free parameters $\{\alpha,\lambda,\kappa\}$ from the empirical data reduces to a two-parameter $\{\alpha,\kappa\}$ fitting problem. To find the parameter values such that the negative of $l(\boldsymbol{\theta};\mathbf{x})$ is minimized, one can use the constrained maximum likelihood estimation method (Schoenberg 1997), which solves the general maximum log-likelihood problem (3.54) subject to the non-linear equality constraint given by λ and bounds $\alpha,\lambda > 0$ and $\kappa \in [0, 1)$.

3.2.1 Description of the Income Data

Data on income distribution were obtained from the Luxembourg Income Study Database (LIS), which provides public access to household-level data files for a large number of countries, including both developed and developing economies.[19] The micro-data are derived from surveys supplied by participating (governmental and non-governmental) statistical agencies at the country level. At the time of writing, LIS includes distributional data on income from 259 national household surveys for 45 countries,[20] covering more than 60 % of the world population (Ravallion 2015). Richer countries are somewhat over-represented within the LIS database, because of data quality and availability. Depending on the country, data are available for different time periods. For the time before 1980, only a small set of countries is found in LIS. By the beginning of the 1980s, the LIS data are organized on the basis of five-year waves until 2000 (Waves I to V) and on a three-year basis since 2004 (Waves VI to IX). The data sets used in this book correspond to all current LIS countries in the most recent years, which are summarily listed in Table 3.1. The sample sizes of the surveys being considered vary from about 2,000 to 230,000 households.

An advantage of using LIS is that income aggregates are defined as uniformly as possible across countries, so that numbers from different countries can be compared more plausibly than using data obtained for each country separately (Gottschalk and Smeeding 2000). For the purposes of this study, we examine the income category referred to as "disposable household income", which is usually the preferred measure for income distribution analysis as it is the income available to households to support their consumption expenditure and saving during the reference period (Canberra Group 2011).[21] According to the LIS documentation (http://www.lisdatacenter.org/data-access/key-figures/disposable-household-income/), this measure includes income received from work, from wealth and from direct government benefits, such as retirement or unemployment benefits. The measure then subtracts direct taxes paid, such as income taxes. The definition does not include sales taxes or non-cash benefits, such as health care provided by a government or employer. The income definition also does not include income from capital gains, a significant source of non-salary income for many wealthy individuals. As a result, many top incomes are probably underestimated.

In all cases, income is expressed in the units of national currency that were in use at the time of data collection. To maintain consistency with the set of methodological practices adopted by the LIS staff to construct the "Inequality

[19] To preserve confidentiality, access to the micro-data is remote, meaning that program code is sent to LIS rather than being run by the user directly on the data.

[20] This is the number of publicly accessible country data sets listed on the LIS website early October 2015 under "LIS List of Datasets" (http://www.lisdatacenter.org/our-data/lis-database/documentation/list-of-datasets/).

[21] The use of different definitions of income is associated with different observed distributions. While the empirical results reported in the following could be sensitive to possible alternative income definitions, this possibility was not explored in the current application. Okamoto (2013) uses LIS data for 25 countries with several income concepts in his analysis—see Chap. 5.

Table 3.1 Income data obtained from the LIS database[a]

Country	Code	Wave IV	Wave V	Wave VI	Wave VII	Wave VIII	Wave IX
Australia	AU					2010	
Austria	AT			2004			
Belgium	BE		2000				
Brazil	BR						2013
Canada	CA					2010	
China	CN		2002				
Colombia	CO					2010	
Czech Republic	CZ					2010	
Denmark	DK					2010	
Egypt	EG						2012
Estonia	EE					2010	
Finland	FI					2010	
France	FR					2010	
Germany	DE					2010	
Georgia	GE						2013
Greece	GR					2010	
Guatemala	GT			2006			
Hungary	HU						2012
Iceland	IS					2010	
India	IN			2004			
Ireland	IE					2010	
Israel	IL						2012
Italy	IT					2010	
Japan	JP				2008		
Luxembourg	LU					2010	
Mexico	MX						2012
Netherlands	NL					2010	
Norway	NO					2010	
Panama	PA						2013
Peru	PE						2013
Poland	PL						2013
Romania	RO	1997					
Russia	RU						2013
Serbia	RS						2013
Slovak Republic	SK					2010	
Slovenia	SI					2010	

(continued)

Table 3.1 (continued)

Country	Code	Wave IV	Wave V	Wave VI	Wave VII	Wave VIII	Wave IX
South Africa	ZA						2012
South Korea	KR			2006			
Spain	ES					2010	
Sweden	SE			2005			
Switzerland	CH			2004			
Taiwan	TW					2010	
United Kingdom	UK					2010	
United States	US						2013
Uruguay	UY						2013

[a]Year given is the income reference year, that is the year to which the income data pertain
Source "LIS List of Datasets" (http://www.lisdatacenter.org/our-data/lis-database/documentation/list-of-datasets/)

and Poverty Key Figures" (http://www.lisdatacenter.org/data-access/key-figures/inequality-and-poverty/), we focus on what is called "equivalized income", i.e. the household income divided by the square root of household size to adjust for differences in household demographics. We also exclude all households where disposable income is missing or exactly equal to zero, and use person-level adjusted weights (the product of household weights and the number of household members) when generating income indicators for the total population. Finally, before proceeding to estimation of income distribution parameters, data have been bottom-coded at 1 % of equivalized mean income and top-coded at 10 times the median of non-equivalized income.

Since the bottom-coding is carried out after equivalization of income (on the equivalized income distribution), while the top-coding is carried out before (on the unequivalized distribution), in the sequel parameter estimation of the κ-generalized and other models for income distribution is performed by solving numerically the general maximum likelihood problem

$$L(\mathbf{x}; \boldsymbol{\theta}) = \prod_{i=1}^{n} \left[f(x_i; \boldsymbol{\theta})^{c_i} F(x_i; \boldsymbol{\theta})^{1-c_i} \right]^{w_i}, \tag{3.58}$$

where the likelihood contribution is given by the density function if the ith observation is not bottom-coded, so that $c_i = 1$, and by the cumulative distribution function for an observation with a bottom-coded value of income, so that $c_i = 0$ (see e.g. Feng et al. 2006; Burkhauser et al. 2011). The Stata command `ml` for user-provided likelihood functions (Gould et al. 2010) is used for estimation of the parameter vector $\boldsymbol{\theta}$.

3.2.2 Results of Fitting to Empirical Distributions: Italy and the United States

Figures 3.8 and 3.9 show the results of fitting the κ-generalized distribution to empirical income data corresponding to the distribution of household incomes in Italy and the United States for the years 2010 and 2013, respectively. The best-fitting parameter values were determined using maximum likelihood estimation as discussed in previous section. This resulted in the following estimates: $\alpha = 2.145 \pm 0.028$, $\beta = 17{,}749 \pm 128$ and $\kappa = 0.622 \pm 0.023$ for Italy; $\alpha = 1.860 \pm 0.010$, $\beta = 38,543 \pm 137$ and $\kappa = 0.625 \pm 0.010$ for the United States. The very small value of the errors indicates that the parameters were precisely estimated, and the comparison between the observed and fitted probabilities in panels (a) and (b) of Figs. 3.8 and 3.9 suggests that the κ-generalized distribution offers a great potential for describing the data over their whole range, from the low to medium income region up to the high income Pareto power-law regime, including the intermediate region for which a clear deviation exists when two different curves are used.

Panel (c) of the same figures depicts the data points for the empirical Lorenz curve, superimposed by the theoretical curve given by Eq. (3.30) with estimates replacing α and κ as necessary. This formula is shown by the solid line in the plots, and fits the data exceptionally well. The plots also compare the empirical Lorenz curve to the theoretical ones associated with the Weibull and Pareto distributions, respectively given by

$$\lim_{\kappa \to 0} L(u) = P\left(1 + \frac{1}{\alpha}, -\ln(1-u)\right), \tag{3.59}$$

where $P(\cdot, \cdot)$ is the lower regularized incomplete gamma function, and

$$\lim_{x \to \infty} L(u) = 1 - (1-u)^{1-\frac{\kappa}{\alpha}}. \tag{3.60}$$

As one can easily recognize, these curves account for only a small part of the whole story.

In order to provide indirect checks on the validity of parameter estimation, we have also calculated the predicted values for the median and mean disposable household income and for two inequality measures like the Gini and Atkinson coefficients—the latter with the inequality aversion parameter ϵ equal to 1. The results of these calculations, performed by substituting the estimated parameters into the relevant expressions given in Sect. 3.1, are shown in Table 3.2 along with their empirical counterparts, which correspond for the considered countries and years to the statistics in the "Inequality and Poverty Key Figures" calculated by LIS staff.[22]

As can be verified, the predictions deduced from the κ-generalized distribution result completely covered by the asymptotic normal 95 % confidence intervals

[22]Computation of inequality measures in this book is performed using the latest version of DASP, the Distributive Analysis Stata Package (Araar and Duclos 2013) available at http://dasp.ecn.ulaval.ca/. The full set of the corresponding "Key Figures" is included in an Excel workbook downloadable at http://www.lisdatacenter.org/data-access/key-figures/download-key-figures/.

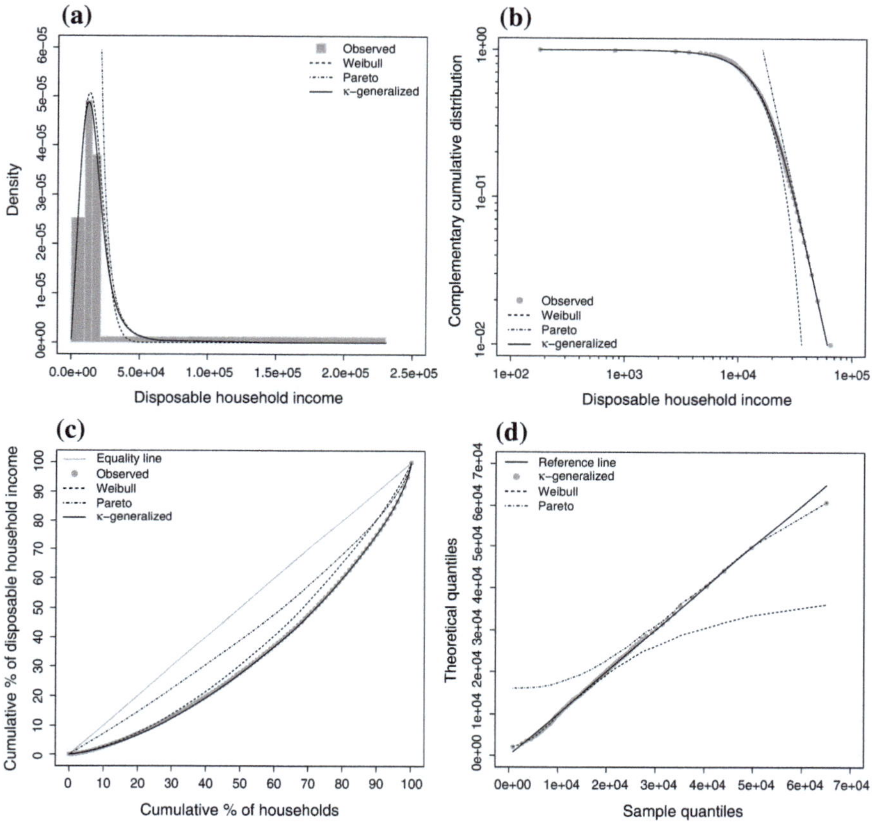

Fig. 3.8 κ-generalized distribution fitted to the 2010 Italian household incomes. **a** Probability density histogram. The *bars*, each representing 1/4 of the total probability, have been drawn after calculating the minimum, the lower quartile, the median, the upper quartile and the maximum of the income variable (e.g. Scott 2015, pp. 73–74). The *solid line* is the κ-generalized model, fitting the data very well in the whole range from the low to the high incomes, including the intermediate income region. This model is compared with the Weibull one (*dashed line*)—fitting the low income data—and with the pure Pareto power-law (*dash-dotted line*)—fitting the high income data. **b** Complementary cumulative distribution. The observed probabilities, corresponding to the empirical percentiles, are plotted on doubly-logarithmic axes, which is the standard way of emphasizing the right-tail behavior of a distribution. Superimposed are the fits of the κ-generalized (*solid line*), Weibull (*dashed line*) and Pareto (*dash-dotted line*) distributions. **c** Lorenz curve. The *filled circles* represent the empirical Lorenz curve corresponding to the percentiles of the distribution, and the *solid line* is the theoretical curve given by Eq. (3.30) using the same parameter values as in panels (**a**) and (**b**). The *gray solid line* corresponds to the Lorenz curve of a society in which everybody receives the same income, and thus serves as a benchmark case against which actual income distribution may be measured. The *dashed and dash-dotted lines* represent the theoretical Lorenz curves from the Weibull and Pareto distributions given by Eqs. (3.59) and (3.60). **d** Q-Q plot of the sample percentiles versus the theoretical percentiles of the fitted κ-generalized (*filled circles*), Weibull (*dashed line*) and Pareto (*dash-dotted line*) distributions. Excellent goodness-of-fit of the κ-generalized model is demonstrated by the fact that the corresponding percentiles lies extremely close to the 45°-ray from the origin, and much closer than is typically observed in plots of this type.

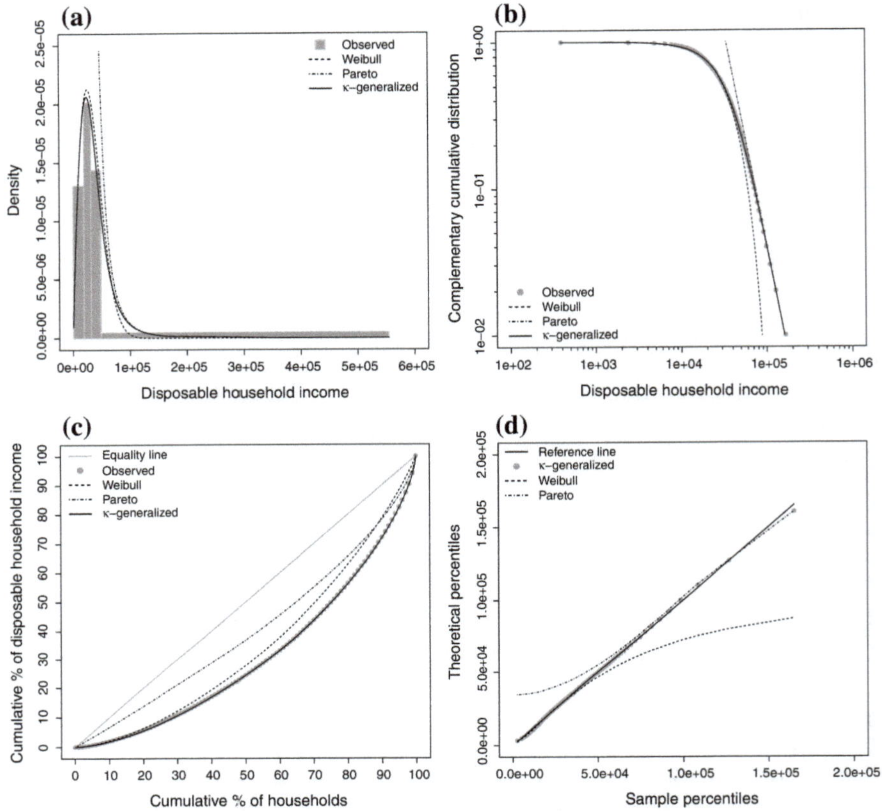

Fig. 3.9 Same plots as in Fig. 3.8 for the US personal income distribution in 2013: **a** probability density; **b** complementary cumulative distribution; **c** Lorenz curve; **d** Q-Q plot

constructed around the empirical values, thus confirming the excellent agreement between the actual model and the sample observations.

The above evidence holds vis-à-vis a further check involving a class of indices that uses standard tools from the economic analysis of income distributions (Bandyopadhyay et al. 2009; Cowell et al. 2013, 2015). Members of this class can be used as measures of discrepancy or goodness-of-fit criteria and have a distributional interpretation that is close to the generalized entropy class of inequality indices.[23] For a sample of size n, the class of considered measures is given by

$$J_\alpha(\mathbf{x}, \mathbf{y}) = \frac{1}{n(\alpha^2 - \alpha)} \sum_{i=1}^{n} \left[\left(\frac{x_i}{\mu_1} \right)^{\alpha} \left(\frac{y_i}{\mu_1} \right)^{1-\alpha} - 1 \right], \tag{3.61}$$

[23]Compared to conventional goodness-of-fit approaches (see for instance D'Agostino and Stephens 1986), the class of fit criteria adopted here is therefore particularly good both at picking up the distinctive shape characteristics of income distribution (heavy-tailedness and right-skewness) and at taking into account inequality and other aspects of social welfare that are usually considered for evaluation of economic performance and policy design.

Table 3.2 Observed and predicted values of the median, the mean, the Gini index G and the Atkinson inequality measure A (1)

Country	Statistic	Observed			Predicted
		Value	LB[a]	UB[b]	
Italy	Median	15,156	14,750	15,562	15,177
	Mean	17,577	17,144	18,009	17,521
	G	0.327	0.317	0.338	0.333
	A (1)	0.189	0.177	0.201	0.183
United States	Median	31,955	31,659	32,250	32,183
	Mean	39,322	38,968	39,676	39,369
	G	0.377	0.374	0.381	0.380
	A (1)	0.234	0.230	0.238	0.236

[a]Lower bound of the 95 % confidence interval
[b]Upper bound of the 95 % confidence interval
Source Authors' calculations based on LIS

where **x** is the sample vector of incomes, **y** the vector of the corresponding quantiles for the theoretical distribution, μ_1 and μ_2 the means of the marginal distributions of **x** and **y**, and $\alpha \in \mathbb{R}$ a sensitivity parameter that can be calibrated according to which part of the income distribution one wants the criterion (3.61) to be particularly sensitive—choosing a large positive value for α would put a lot of weight on discrepancies between a proposed model of income distribution and the data in the upper tail, whereas choosing a substantial negative value would put more weight on lower-tail discrepancies.[24] To assess the suitability of the κ-generalized model for the Italian and US income data, we interpret the class of measures J_α as goodness-of-fit measures by testing the hypothesis $H_0 : J_\alpha(\mathbf{x}, \mathbf{y}) = 0$ against the alternative $H_1 : J_\alpha(\mathbf{x}, \mathbf{y}) \neq 0$, where **y** is the n-sized vector of quantiles $y_i = F^{-1}\left(\frac{i}{n+1}; \hat{\alpha}, \hat{\beta}, \hat{\kappa}\right)$ derived from the κ-generalized parameter estimates.[25] Table 3.3 reports the estimated values of J_α, for $\alpha = -1, 0, 1, 2$, along with the associated p-values obtained by means of 250 Monte Carlo simulations.[26] Observe that the κ-generalized distribution provides a

[24]Like the generalized entropy class of inequality measures, expression (3.61) is not defined for $\alpha = 0$ and $\alpha = 1$, as the denominator $n\left(\alpha^2 - \alpha\right) = 0$ in both cases. Expressions for these values of α (the "middle-sensitive" goodness-of-fit criteria) are therefore calculated by using L'Hôpital rule, by which the limit of an undefined ratio between two functions of the same variable is equal to the limit of the ratio of their first derivatives. Expressions for each J_α index other than for the cases $\alpha = 0, 1$ can be, instead, derived by substitution.

[25]Note that we use $\frac{i}{n+1}$, rather than $\frac{i}{n}$, as an estimate of the empirical distribution function. Had we used $\frac{i}{n}$, then the value of the empirical cumulative would automatically be set to 1 when $i = n$ and the inversion of F would return an infinite value—see e.g. expression (3.21).

[26]In detail, our procedure is as follows. First, we calculate the test statistics for the κ-generalized model fitted to the empirical data. Next, we generate 250 synthetic data sets from the κ-generalized distribution where unknown parameters are replaced by their sample estimates. We then fit each synthetic data set individually to the κ-generalized distribution and calculate the test statistics for each one relative to its own model. Finally, we count what fraction of the time each resulting

Table 3.3 The κ-generalized model for Italian and US income distributions: J_α goodness-of-fit tests[a]

Country	α	$J_\alpha \times 10^2$	
		Index	p-value[b]
Italy	–1	0.183	0.028*
	0	0.122	0.064
	1	0.098	0.100
	2	0.087	0.120
United States	–1	0.089	0.040*
	0	0.077	0.088
	1	0.068	0.112
	2	0.063	0.159

[a]The null hypothesis is that the discrepancy between the fitted model and the data is zero
[b]Computed via 250 Monte Carlo simulations; $*$ = significant at the 5 % level
Source Authors' calculations based on LIS

satisfactory fit to the data of both countries (small values of the test statistics) independently on the part of the distribution one is interested in testing for goodness-of-fit, but the discrepancy with the empirical data at low incomes is statistically significant at the 5 % level. However, if one lowers the significance level at 1 %, H_0 is no more rejected for $\alpha = -1$, so for a "bottom-sensitive" goodness-of-fit criterion the κ-generalized would still constitute a satisfactory fit in both situations. In particular, the discrepancy at the upper end of the income distribution between the empirical data and the theoretical model—as assessed by the "top-sensitive" goodness-of-fit criterion J_2—is so small that the p-value is not able to lead to rejection of the null hypothesis at any of the usual significance levels (1, 5 and 10 %).

The linear behavior emerging from the quantile-quantile (Q-Q) plot of the sample percentiles versus the corresponding percentiles of the fitted κ-generalized distribution and its two limiting cases displayed in panel (d) of Figs. 3.8 and 3.9 confirms the quantitative results obtained by hypothesis testing, as well as the fact that the Weibull and Pareto distributions can give only a partial and incomplete description of the data.

(Footnote 26 continued)
statistic is larger than the value for the empirical data. This fraction is the p-value for each fit, and can be interpreted in the standard way: if it is larger than the chosen significance level, then the difference between the empirical data and the model can be attributed to statistical fluctuations alone; if it is smaller, the model is not a plausible fit to the data. Note crucially that for each synthetic data set we compute the test statistics relative to the best-fit model for *that* data set, not relative to the distributions fitted to the original data. In this way we ensure that we are performing for each synthetic data set the same calculations we performed for the real data, a crucial requirement if we wish to get unbiased estimates of the p-values (see e.g. Capasso et al. 2009).

3.2.3 Comparisons of Alternative Distributions

In this section, comparisons of the performance of the κ-generalized distribution with other parametric models proposed in the literature are presented. Namely, these models are the three-parameter Singh-Maddala (Singh and Maddala 1976) and Dagum type I (Dagum 1977) distributions, which have, respectively, the density

$$f(x; a, b, q) = \frac{aqx^{a-1}}{b^a \left[1 + (x/b)^a\right]^{1+q}}, \quad x > 0, \quad a, b, q > 0, \tag{3.62}$$

and

$$f(x; a, b, p) = \frac{apx^{ap-1}}{b^{ap} \left[1 + (x/b)^a\right]^{p+1}}, \quad x > 0, \quad a, b, p > 0. \tag{3.63}$$

Kleiber and Kotz (2003, Chap. 6) report analytical expressions for the distribution functions, moments and several tools for inequality measurement—included the Lorenz curve and Gini coefficient—corresponding to (3.62) and (3.63). More recently, Jenkins (2009) provided formulas for the generalized entropy measures of the GB2 distribution, from which the Singh-Maddala and Dagum versions are also easily obtained.[27]

Our comparisons involve all LIS countries listed in Table 3.1—a total of 45 data sets. Table 3.4 provides distributional summary measures for disposable household income in each country and year.[28] The maximum likelihood estimates for the three models under consideration are presented in Table B.1. It is observed that the κ-generalized gives the best results in the sense that the parameter standard errors—derived from the inverse of the Hessian matrix obtained after convergence of the log-likelihood was reached—are generally the lowest among the competing models of income distribution.[29]

In order to judge which distribution better models the data, we use the J_α indices interpreted as measures of discrepancy (see previous section). The results are shown in Table B.2. In general, the κ-generalized has the flexibility to provide a good fit

---

[27]Let X be a random variable following the GB2 distribution with parameters a, b, p and q, that is $X \sim \text{GB2}(a, b, p, q)$. The Singh-Maddala distribution is the special case of the GB2 distribution when $p = 1$; the Dagum type I distribution is the special case when $q = 1$. For a discussion of other special cases, see McDonald (1984) and Kleiber and Kotz (2003).

[28]For convenience of presentation, each data set is labeled with an abbreviation concatenating the two-digit ISO country code with the last two digits of the reference year. The country and year four-digit codes can be found in the LIS "Datasets Information" (http://www.lisdatacenter.org/our-data/lis-database/datasets-information/).

[29]In the cases of South Africa (ZA12) and Uruguay (UY13), the maximum likelihood estimate of the tail parameter κ slightly exceeds the upper bound of 1 for the parameter space. Since micro-data on individual observations cannot be downloaded, but only remotely accessed through the LIS server, it was not possible for us to determine whether this is due to significant heaviness of the upper tails or to the presence of questionable values for income that greatly affected the approximation of the κ-generalized function in the upper tail region of these distributions.

Table 3.4 Distributional summary statistics on disposable household income: all LIS countries in the most recent years[a]

Code[b]	Observations	Mean	G	MLD	$GE(2)$	$A(1)$	S
AU10	17,968	49,439	0.330	0.202	0.242	0.183	9.221
AT04	5,147	22,556	0.269	0.128	0.157	0.120	8.408
BE00	2,080	748,510	0.279	0.139	0.227	0.130	7.463
BR13	110,888	19,145	0.450	0.362	0.541	0.304	11.394
CA10	25,019	41,988	0.317	0.186	0.235	0.169	9.522
CN02	17,108	7,746	0.505	0.498	0.563	0.392	9.586
CO10	16,006	7,087,289	0.483	0.439	0.619	0.355	8.876
CZ10	8,864	245,799	0.256	0.112	0.141	0.106	8.807
DK10	85,645	247,152	0.248	0.125	0.152	0.117	11.203
EG12	11,849	10,624	0.464	0.392	0.599	0.324	9.088
EE10	4,985	116,863	0.325	0.192	0.205	0.175	8.255
FI10	9,348	26,455	0.261	0.118	0.164	0.111	8.901
FR10	15,680	23,882	0.289	0.155	0.200	0.143	9.174
DE10	12,137	22,858	0.286	0.143	0.187	0.133	9.026
GE13	2,758	4,485	0.394	0.268	0.352	0.235	7.695
GR10	5,993	15,170	0.333	0.216	0.269	0.194	8.242
GT06	13,644	20,089	0.490	0.446	0.625	0.360	8.984
HU12	2,054	1,634,275	0.289	0.151	0.184	0.140	7.416
IS10	3,015	3,884,131	0.245	0.110	0.156	0.104	7.948
IN04	41,440	20,712	0.491	0.471	0.587	0.376	10.214
IE10	4,303	25,380	0.294	0.159	0.172	0.147	7.865
IL12	8,742	94,466	0.371	0.259	0.288	0.228	8.835
IT10	7,904	17,577	0.327	0.209	0.238	0.189	8.428
JP08	3,360	3,328,014	0.302	0.167	0.196	0.154	8.022
LU10	5,464	41,191	0.269	0.127	0.155	0.120	8.042
MX12	8,998	68,323	0.459	0.399	0.559	0.329	8.684
NL10	10,492	24,940	0.257	0.124	0.147	0.117	8.945
NO10	227,190	339,518	0.243	0.123	0.135	0.116	12.166
PA13	11,790	6,983	0.467	0.400	0.555	0.330	9.015
PE13	29,970	13,716	0.455	0.437	0.480	0.354	9.890
PL13	37,086	27,368	0.316	0.195	0.235	0.177	10.307
RO97	32,187	8,164,784	0.280	0.136	0.196	0.127	10.137
RU13	5,658	304,486	0.331	0.197	0.241	0.178	8.522
RS13	4,471	391,860	0.332	0.232	0.245	0.207	8.224
SK10	5,198	8,387	0.263	0.128	0.154	0.120	8.402
SI10	3,923	13,645	0.252	0.126	0.115	0.119	8.174
ZA12	7,990	38,907	0.572	0.623	0.895	0.463	8.351
KR06	15,514	18,587,582	0.311	0.191	0.189	0.173	9.217

(continued)

Table 3.4 (continued)

Code[b]	Observations	Mean	G	MLD	GE (2)	A (1)	S
ES10	12,786	16,249	0.334	0.237	0.202	0.211	9.089
SE05	16,268	203,460	0.237	0.102	0.126	0.097	9.568
CH04	3,270	50,020	0.268	0.147	0.157	0.137	7.905
TW10	14,853	450,158	0.318	0.178	0.222	0.163	9.458
UK10	25,275	20,233	0.335	0.212	0.295	0.191	9.900
US13	51,105	39,322	0.377	0.267	0.341	0.234	10.557
UY13	46,582	253,528	0.372	0.233	0.324	0.208	10.539

[a]G = Gini; MLD = mean logarithmic deviation; GE (2) = top-sensitive generalized entropy index; A (1) = Atkinson index with inequality aversion parameter $\epsilon = 1$; S = Shannon entropy
[b]See footnote 28
Source Authors' calculations based on LIS

to different parts of the income distributions, although it is not definitively the best of the three-parameter models in all cases. Nonetheless, the κ-generalized is able to outperform the other two models for the majority of the country data sets analyzed. Looking indeed at the relative ranking of the measures considered, the discrepancy between κ-generalized and data at low incomes ($\alpha = -1$) is the smallest in 40 % of the cases,[30] whereas for the Singh-Maddala and Dagum type I distributions is so in about 29 % and 31 % of the cases, respectively. The approximation of the κ-generalized to real data is also the closest for the middle range of incomes in no less than 42 % ($\alpha = 0$) and 48 % ($\alpha = 1$) of the cases[31]—as against 31 % and 23 %, which are the average percentages of country data sets that are best modeled by the Singh-Maddala and Dagum type I distributions according to both the "middle-sensitive" discrepancy measures. Finally, when focusing on incomes in the upper tail ($\alpha = 2$), the κ-generalized model still offers superior performance over its two competitors in almost 49 % of the cases,[32] while it is outperformed by the estimated Singh-Maddala and Dagum type I distributions in, respectively, 33 % and 18 % of the data sets.

[30]Corresponding to the income distributions of Austria (AT04), Belgium (BE00), Finland (FI10), Greece (GR10), Hungary (HU12), Iceland (IS10), Israel (IL12), Italy (IT10), Japan (JP08), Luxembourg (LU10), Netherlands (NL10), Peru (PE13), Romania (RO97), Slovak Republic (SK10), Spain (ES10), Sweden (SE05), United Kingdom (UK10) and the United States (US13).

[31]Corresponding all at once to the income distributions of Austria (AT04), Belgium (BE00), Canada (CA10), Denmark (DK10), Finland (FI10), Greece (GR10), Hungary (HU12), Iceland (IS10), Israel (IL12), Italy (IT10), Japan (JP08), Netherlands (NL10), Norway (NO10), Peru (PE13), Poland (PL13), Romania (RO97), Serbia (RS13), Slovak Republic (SK10), Sweden (SE05), Switzerland (CH04), United Kingdom (UK10) and the United States (US13).

[32]Corresponding to the income distributions of Austria (AT04), Belgium (BE00), Canada (CA10), Denmark (DK10), Finland (FI10), Greece (GR10), Hungary (HU12), Iceland (IS10), Israel (IL12), Italy (IT10), Japan (JP08), Netherlands (NL10), Norway (NO10), Peru (PE13), Poland (PL13), Romania (RO97), Serbia (RS13), Slovak Republic (SK10), Sweden (SE05), Switzerland (CH04), United Kingdom (UK10) and the United States (US13).

Since a "good" model of income distribution should not only capture the shape of empirical distributions but also be close to them in a sense that is consistent with considerations of economic welfare, the performance of the three models under scrutiny is further evaluated by considering the accuracy of selected distributional statistics implied by parameter estimates. To this end, Table B.3 reports the predicted values for the mean and several inequality measures deduced from the estimated distributional parameters presented in Table B.1.[33] For any of the models examined, the accuracy of implied statistics is assessed by calculating the absolute percentage error

$$APE = \frac{|P - A|}{A} \times 100 \tag{3.64}$$

between the predicted values (P) and the actual sample estimates (A) given in Table 3.4. The results are summarized in Fig. 3.10, which shows the percentage of cases where each model attains the lowest APE for the distributional statistics being considered. The closeness between the implied and sample estimates of the mean is the most recurrent for the κ-generalized distribution than for the Singh-Maddala and Dagum type I. Conversely, the implied estimates of the Gini coefficient (G) are frequently more accurate for the latter models than for the competing κ-generalized. Hence the estimation procedure of the κ-generalized parameters appears to have roughly preserved the mean characteristic for most of the country data sets analyzed, but in many cases it implicitly modeled intra- and/or inter-group variation incorrectly. However, when looking at income differences in different parts of the income distribution, the κ-generalized provides relatively more accurate estimates than its competitors of the mean logarithmic deviation (MLD), the top-sensitive $GE\,(2)$ index and the Atkinson inequality measure $A\,(1)$. Recalling that the MLD index is relatively sensitive to middle-income differences, while the indices $GE\,(2)$ and $A\,(1)$ are more sensitive to what happens respectively at the upper and bottom ends of the income distribution, the above results provide consistent evidence with the closest approximation to the income distribution shape found for the κ-generalized model in a significant number of cases.

Finally, the agreement between the informational entropy of the observed data and the entropy of the fitted κ-generalized distribution is the closest in more than 80 % of the cases analyzed, whereas the percentages for the Singh-Maddala and Dagum type I models are lower than 9 % (see the last group of bars in Fig. 3.10).[34] The empirical entropy has been obtained by replacing the discrete probabilities f_i, $i = 1, \dots, m$,

---

[33]Since the top-sensitive $GE\,(2)$ index is a monotonic transformation of the coefficient of variation, calculation of this inequality measure has not been possible in some cases due to non-existence of the variance, which happens for the Singh-Maddala and Dagum type I distributions when, respectively, $aq < 2$ and $a < 2$ (Kleiber and Kotz 2003, Chap. 6) and for the κ-generalized when $\frac{\alpha}{\kappa} < 2$ (see Sect. 3.1.3). The cases affected correspond to the income distributions of Brazil (BR13), China (CN02), Colombia (CO10), Egypt (EG12), Guatemala (GT06), India (IN04) and South Africa (ZA12), and are indicated in the table by a dash ("–") symbol.

[34]Comparing the entropy of the data to the entropy based on a particular probability distribution can provide relevant insights into the appropriateness of the distributional model given the data.

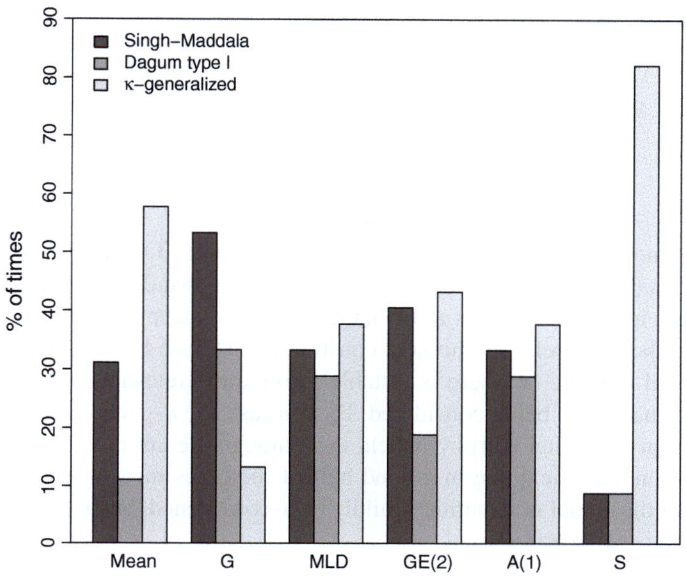

Fig. 3.10 Percentage of times in which the Singh-Maddala, Dagum type I and κ-generalized models attain the lowest absolute percentage error between the predicted values for key distributional summary measures and their sample counterparts. Countries affected by non-existence of the variance of one or more than one of the models considered are excluded from calculation of the percentages for the top-sensitive generalized entropy index $GE(2)$—see footnote 33

in the Shannon entropy measure $S = -\sum_{i=1}^{m} f_i \ln(f_i)$ by maximum likelihood estimates $\hat{f}_i = \frac{n_i}{n} = \frac{n_i}{\sum_{i=1}^{m} n_i}, n \geq m$, where n_i is the weighted frequency of realization i in the sample of n observations. The corresponding analytical expressions, derived by solving formula (3.4) in terms of the relevant probability density functions, are given by

$$S = -\ln(aq) + \ln(b) + (a-1)\frac{\psi(q)+\gamma}{a} + \frac{1}{q} + 1, \tag{3.65}$$

for the Singh-Maddala,

$$S = -\ln(ap) + \ln(b) + (1-ap)\frac{\psi(p)+\gamma}{a} + (p+1)\left[\psi(p+1)+\gamma\right], \tag{3.66}$$

(Footnote 34 continued)
If the entropy of the data is lower than the entropy based on an assumed distribution function, then there is information in the data that is not being taken advantage of to specify the distribution function. On the other hand, if the entropy of the data is greater than that based on the hypothetical distribution function, more structure is being assumed than is justified by the data. See Jaynes (1978) for a more thorough discussion of this approach and Schneider (2008, 2015) for an empirical application to US earned incomes.

for the Dagum type I, and

$$
\begin{aligned}
S = & -\ln\left(2\alpha\kappa^{\frac{1}{\alpha}-1}\right) + \ln(\beta) + (\alpha - 1)\frac{\psi\left(\frac{1}{2\kappa}\right) + \kappa + \gamma}{\alpha} \\
& + \frac{1}{2}\left[\psi\left(\frac{1}{2} + \frac{1}{4\kappa}\right) - \psi\left(1 + \frac{1}{4\kappa}\right)\right] + \kappa - 1,
\end{aligned}
\tag{3.67}
$$

for the κ-generalized. An analysis of the entries in Tables 3.4 and B.3 suggests that the entropies (3.65) and (3.66) deduced from the Singh-Maddala and Dagum type I distributions are in many cases much larger than the entropy of the data. This fact means that there are relevant moment constraints that have been neglected in these distributional models, i.e. there is testable information relevant to the description of the data that is not being considered. By contrast, the κ-generalized distribution is found to give a better entropy match, even though the entropy (3.67) is slightly lower than the empirical one in around half of the cases investigated. This would suggest that the model is sometimes mildly over-constrained, but at least no relevant information is being left unused.

Chapter 4
The κ-Generalized Mixture Model for the Size Distribution of Wealth

Wealth—any income that is at least one hundred dollars more a year than the income of one's wife's sister's husband.
Henry Louis Mencken

Abstract The chapter illustrates the main analytical properties of the κ-generalized net wealth distribution model, which is most able to accommodate the special features of wealth data. Negative, zero and positive data are modeled with a Weibull distribution, a point-mass at zero and a κ-generalized distribution, respectively. Expressions for the mean and popular tools for analyzing inequality are also derived for the assumed model of net wealth distribution. Finally, the specified model is fitted to a number of country data sets on household net wealth and its performance compared to alternative mixture models using the Dagum type I and Singh-Maddala specifications as descriptions of the positive net wealth values.

Keywords Net wealth distribution · Parametric modeling · Finite mixture models · κ-generalized mixture model

4.1 Motivation

In addition to income, wealth is an important indicator of economic well-being and is largely used to assess the living standards of individuals, families or households. Both income and wealth may be used to finance current consumption or retained to support future consumption, and can thus be thought of as alternative means of securing a standard of living. Clearly, some correlation exists between income and wealth, but the relationship between these two variables is not perfect: greater income is in fact likely to mean greater wealth, but not always.[1] This occurs because the two

[1]For instance, a high income may be associated with low wealth—this is generally the case with young people starting their careers; on the other hand, a low income could accompany high wealth—this is the case with some retirees who have little income but have accumulated and paid

F. Clementi and M. Gallegati, *The Distribution of Income and Wealth*, New Economic Windows, DOI 10.1007/978-3-319-27410-2_4

concepts are not synonymous. Income is a *flow*, since it is meaningful only when defined in relation to a period of time (hourly, weekly, monthly or annual income). Wealth is a *stock*, increasing as new assets are acquired or savings accumulated, and the only time information required is when the stock was measured (no periodicity is necessary). The link between the flow from income and the stock of wealth is obvious: the greater the former, the more rapidly the latter increases. More generally, the higher the correlation between income and wealth, the higher is the level of potential consumption due to one or the other variable. There can be therefore little dispute that wealth is a relevant measure of living standard too, and one which is probably able to capture long-term economic resources better than income—as, over and above any income flow, it represents resources that people are able to draw upon to face adverse shocks.

At the practical level, wealth distribution varies widely both across countries and over time, but it is always more unequal than income distribution due to the very high levels of concentration among the extremely wealthy (Cowell and Van Kerm 2015; Davies et al. 2009, 2011; Roine and Waldenström 2015). Life-cycle savings and bequest motives feature prominently in the set of factors that have a large influence in shaping wealth inequality and its concentration at the top (Piketty and Zucman 2015). In addition, wealth inequality is much larger than income inequality due to financial assets that are very unequally distributed and mainly accrue to top income and top wealth households (Davies and Shorrocks 2000; Murtin and Mira d'Ercole 2015; OECD 2015). The distinctive features of empirical wealth distributions, however, make measurement of inequality somewhat more challenging than for the analysis of income distributions. These features include the presence of negative and zero data on wealth as well as the positive skewness with "fat" and long right-hand tails, which result in sparse and extreme upper wealth data. The presence of negative values of wealth, in particular, has been documented to make some traditional measures of inequality inadequate (Amiel et al. 1996; Chen et al. 1982; Cowell 2011; Cowell and Van Kerm 2015; Hagerbaumer 1977; Jenkins and Jäntti 2005; Pyatt et al. 1980; Stich 1996).

The singularities of wealth distribution requires at same time the design of different parametric models from those typically used for income distribution. Indeed, the problem for the wealth researcher is that virtually all of the models suggested within the context of the income distribution literature are defined for variables taking only strictly positive values, although published statistical data of wealth distributions give clear evidence of presenting highly significant frequencies of households or individuals with null and/or negative wealth. The early contributions systematically dismissed these frequencies and fitted their respective proposed models to the positive observations only, thus omitting a significant part of the story.[2]

(Footnote 1 continued)
for substantial assets that could be drawn upon in order to supplement their income and maintain the desired level of consumption.

[2]In the 1950s, Wold and Whittle (1957) and Sargan (1957) proposed the Pareto type I model and the lognormal distribution, respectively. Afterward, other models were proposed: in 1969 the Pareto types I and II by Stiglitz (1969); in 1975, the log-logistic by Atkinson (1975) and the Pearson type V

Dagum (1977, 1978) was the first to specify and test a four-parameter model for wealth distributions (Dagum type II). The fourth parameter in the Dagum model is an estimate of the frequency of economic units with wealth equal to zero. This model is highly relevant to describe total (gross) wealth distribution because of the always large observed percentage of economic units with null total wealth. Dagum (1990, 1994, 2006a, b) made further developments of his type II model to analyze the distribution of *net* wealth, i.e. the value of gross wealth minus total debt, which is considered by many researchers the relevant wealth concept for empirical analysis. The support of the Dagum model of net wealth is the real line $\mathbb{R} = (-\infty, \infty)$, thus allowing to fit the subset of economic units with null and negative wealth. Furthermore, it contains as particular cases both the Dagum types I and II distributions (Dagum 1977).

More in detail, the Dagum general model of net wealth distribution is a mixture (or a convex combination) of an atomic and two continuous distributions. The atomic distribution concentrates its unit mass of economic agents at zero, and therefore accounts for the economic units with null net wealth. The continuous distribution accounting for the negative net wealth observations is given by a Weibull function. It has a fast left tail convergence to zero, and therefore it has finite moments of all orders. The other continuous distribution, specified as the Dagum type I model, accounts for the positive values of net wealth and presents a heavy right tail, thus having a small number of finite moments of positive order. This different behavior at the two tails of the distribution stems form the fact that, unlike the right tail of income and (gross or net) wealth distributions—which tend slowly to zero when income and wealth tend to infinity, the distribution of the negative values (left tail) of net wealth tends very fast to zero when the variable tends to minus infinity, since economic units face a short-term challenge of either moving out of the negative range of net wealth or bankruptcy.

The Dagum model of net wealth distribution allows comprehensive description of the overall distribution, including the negative values and the spike at zero that is typically observed in most sample data on wealth. However, relatively few applications of this model other than Dagum (1990, 2006a) ones to Italian wealth data are known at the time of writing—the only notable exceptions being represented by Jenkins and Jäntti (2005), who fitted the model to Finnish net wealth data in 1984 and 1989, and by Cowell and Van Kerm (2015), who provide an application to 15 euro area countries based on the 2013 wave of the Household Finance and Consumption Survey (HFCS), a joint project of all central banks of the Eurosystem. Departures from Dagum original specification are also very uncommon among parametric models of net wealth distribution. The only attempts that we are aware of in the current literature were Jäntti et al. (2015), who use a Singh-Maddala distribution for positive data instead of a Dagum type I, and Clementi et al. (2012b), who replace the Dagum type I model with the κ-generalized functional form—the latter authors also highlight the relative merits of their specification with respect to finite mixture models based

(Footnote 2 continued)
by Vaughan (1975). All of these models are restricted to describe only the positive range of wealth, since they are not defined for zero and/or negative values.

upon the Dagum type I and Singh-Maddala distributions for the positive values of 1984–2011 US net wealth data.

The purpose of the present chapter is to provide additional evidence in support of the use of finite mixture models for their ability to handle the specificity of wealth distributions, in particular the presence of zero and negative values. Motivated by our previous results (Clementi et al. 2012b), the focus will be primarily on illustrating the main analytical properties of the κ-generalized net wealth distribution model, where negative, zero and positive data are modeled respectively with a Weibull distribution, a point-mass at zero and a κ-generalized distribution (Sect. 4.2). Expressions for the mean (Sect. 4.3) and popular tools for analyzing inequality (Sect. 4.4) will then be derived for the assumed model of net wealth distribution. Finally, the specified model will be fitted to a number of country data sets on household net wealth included in the Luxembourg Wealth Study (LWS) database and its performance compared to alternative mixture models using the Dagum type I and Singh-Maddala specifications as descriptions of the positive net wealth values (Sect. 4.5).

4.2 Model Specification

The κ-generalized model of net wealth distribution as a mixture of an atomic and two continuous distributions takes the form

$$f(w) = \sum_{i=1}^{3} \theta_i f_i(w), \quad -\infty < w < \infty, \quad \theta_i \geq 0, \ \sum_i \theta_i = 1, \tag{4.1}$$

where w denotes the wealth variable and $\theta_i, i = 1, \ldots, 3$, are the mixture proportions. The two-parameter Weibull density

$$f_1(w) = \frac{s}{\lambda}\left(\frac{|w|}{\lambda}\right)^{s-1} \exp\left[-\left(\frac{|w|}{\lambda}\right)^{s}\right], \quad w < 0, \qquad s, \lambda > 0, \tag{4.2}$$

describes the distribution of economic units with negative net wealth, while the null net wealth observations are accounted for by a distribution that concentrates its unit mass at $w = 0$, i.e.

$$f_2(0) = 1. \tag{4.3}$$

The other continuous distribution, $f_3(w)$, accounts for the positive values of net wealth and is specified by the three-parameter κ-generalized density

$$f_3(w) = \frac{\alpha}{\beta}\left(\frac{w}{\beta}\right)^{\alpha-1} \frac{\exp_\kappa\left[-(w/\beta)^\alpha\right]}{\sqrt{1+\kappa^2 (w/\beta)^{2\alpha}}}, \quad w > 0, \quad \alpha, \beta > 0, \quad \kappa \in [0, 1). \tag{4.4}$$

The corresponding cumulative distribution function reads

$$F(w) = \theta_1 F_1(w) + \theta_2 F_2(w) + \theta_3 F_3(w), \tag{4.5}$$

with

$$F_1(w) = \begin{cases} \exp\left[-\left(\frac{|w|}{\lambda}\right)^s\right] & \text{if } w < 0, \\ 1 & \text{if } w \geq 0; \end{cases} \tag{4.6a}$$

$$F_2(w) = \begin{cases} 0 & \text{if } w < 0, \\ 1 & \text{if } w \geq 0; \end{cases} \tag{4.6b}$$

$$F_3(w) = \begin{cases} 0 & \text{if } w \leq 0, \\ 1 - \exp_\kappa\left[-(w/\beta)^\alpha\right] & \text{if } w > 0. \end{cases} \tag{4.6c}$$

It follows easily that

$$F(w) = \begin{cases} \theta_1 \exp\left[-\left(\frac{|w|}{\lambda}\right)^s\right] & \text{if } w < 0, \\ \rho & \text{if } w = 0, \\ \rho + (1-\rho)\left\{1 - \exp_\kappa\left[-(w/\beta)^\alpha\right]\right\} & \text{if } w > 0, \end{cases} \tag{4.7}$$

where $\rho = \theta_1 + \theta_2$ and $1 - \rho = \theta_3$. It can be verified that when $\theta_1 = \theta_2 = 0$ (hence $\rho = 0$) and $\theta_3 = 1 - \rho = 1$, the κ-generalized distribution function (3.16) is recovered.

4.3 Moments of the κ-Generalized Mixture Model for Net Wealth Distribution

From (4.1), the rth-order moment about the origin is

$$\mu_r^{'} = E\left(W^r\right) = \int_{-\infty}^{\infty} w^r f(w)\, \mathrm{d}w = \theta_1 E_1\left(W^r\right) + \theta_2 E_2\left(W^r\right) + \theta_3 E_3\left(W^r\right), \tag{4.8}$$

with

$$E_1\left(W^r\right) = (-1)^r \lambda^r \Gamma\left(1 + \frac{r}{s}\right), \tag{4.9a}$$

$$E_2\left(W^r\right) = 0, \tag{4.9b}$$

$$E_3\left(W^r\right) = \beta^r (2\kappa)^{-\frac{r}{\alpha}} \frac{\Gamma\left(1 + \frac{r}{\alpha}\right)}{1 + \frac{r}{\alpha}\kappa} \frac{\Gamma\left(\frac{1}{2\kappa} - \frac{r}{2\alpha}\right)}{\Gamma\left(\frac{1}{2\kappa} + \frac{r}{2\alpha}\right)}. \tag{4.9c}$$

Specifically, the mean net wealth equals

$$\mu_1^{'} = E(W) = m = -\theta_1 \lambda \Gamma \left(1 + \frac{1}{s}\right) + \theta_3 E_3(W), \tag{4.10}$$

where $E_3(W)$ is given by Eq. (4.9c) with $r = 1$ [compare with (3.24)].

4.4 The Lorenz Curve and the Gini Index of the Net Wealth Distribution Model

Given the mathematical structure of the net wealth distribution function specified by (4.7), we have

$$L(u) = \begin{cases} -\frac{\lambda\theta_1}{m}\Gamma\left(1+\frac{1}{s}, \log\frac{\theta_1}{u}\right) & \text{if } 0 \le u < \theta_1, \\ -\frac{\lambda\theta_1}{m}\Gamma\left(1+\frac{1}{s}\right) & \text{if } \theta_1 \le u \le \rho, \\ \frac{1}{m}\left\{\frac{(1-\rho)\beta}{(2\kappa)^{1+\frac{1}{\alpha}}}\left[B\left(\frac{1}{2\kappa}-\frac{1}{2\alpha}, 1+\frac{1}{\alpha}\right)\right.\right. & \\ \left.-B\left(\left[\frac{1-u}{1-\rho}\right]^{2\kappa}; \frac{1}{2\kappa}-\frac{1}{2\alpha}, 1+\frac{1}{\alpha}\right)\right] & \text{if } u > \rho, \\ \left.-\lambda\theta_1\Gamma\left(1+\frac{1}{s}\right)\right\} & \end{cases} \tag{4.11}$$

with $L(1) = 1$. Equation (4.11) determines the path of the Lorenz curve for the κ-generalized mixture model over the whole range of wealth. A graphical representation of its shape is given in Fig. 4.1. Assuming mean wealth for the whole population is positive, as in panel (a) of the figure, the first part of $L(u)$ drops below the horizontal line at zero and has a negative slope for negative net wealth values; then the curve is horizontal, corresponding to households that have zero net wealth, and has the conventional positive slope over the remaining units (with positive wealth values). Relative to the conventional picture of a Lorenz curve, the curve (4.11) takes on an even more non-standard shape when $m < 0$, as in this case it appears to be flipped vertically—see panel (b) of the figure. Indeed, starting from the poorest unit, the curve has a positive slope and lies above the zero line over the range of negative wealth values; then it is horizontal where wealth is zero, and has a negative slope over the remaining (positive) wealth units.

Since the shape and position of the net wealth Lorenz curve are not always the ones that conventionally arise in the analysis of incomes, because of the high incidence of negative and zero values in wealth data, alternative representational devices such as the Generalized Lorenz curve (Shorrocks 1983) might have some advantages in this

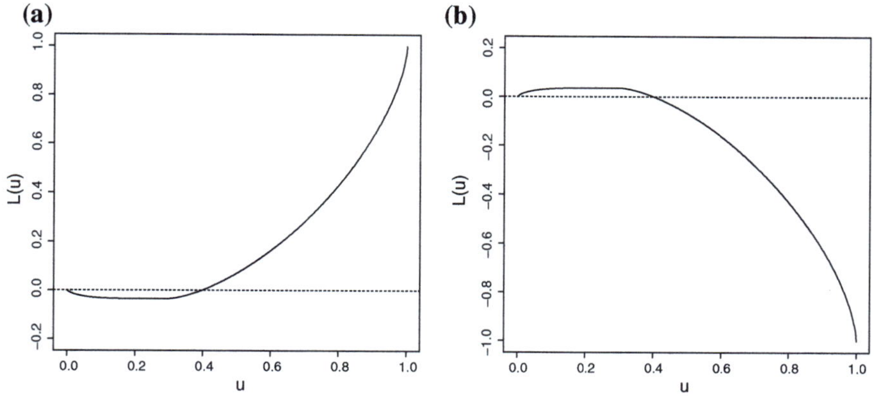

Fig. 4.1 The Lorenz curve of the κ-generalized mixture model for net wealth distribution. Both curves in panels (**a**) and (**b**) have been drawn using the following parameter values: $s = 0.7, \alpha = 2$, $\beta = 10$, $\kappa = 0.75$, $\lambda = 1$, $\theta_1 = 0.2$ and $\theta_2 = 0.1$. Mean net wealth m equals, respectively, 7.172 and −7.172

context. The Generalized Lorenz curve depicts the cumulative wealth of a population scaled up by mean wealth at each point, i.e.

$$GL(u) = mL(u). \tag{4.12}$$

The curve (4.12) is well-defined for all values of wealth along the real line, and its slope at each wealth value w is equal to w itself. Specifically, the curve is negatively sloped and below the zero line as long as wealth is negative, horizontal at the points where nil wealth values are encountered, and positively sloped above zero over the rest of the diagram (i.e. for units with positive net wealth). The right-hand intercept of $GL(u)$ is m, and thus the Generalized Lorenz curve is not affected in the way the conventional Lorenz curve is when $m < 0$—it lies completely below the zero line in this case, but the curve is no more flipped vertically. Figure 4.2 illustrates the shape of $GL(u)$ for the κ-generalized mixture model of net wealth distribution derived from the joint applications of Eqs. (4.11) and (4.12). The values of the parameters and mean net wealth m are the same as in Fig. 4.1a.

Following from the fact that the net wealth Lorenz curve (4.11) presents negative values for all $u < \rho$, it can be proved that the Gini index takes the general form (Dagum 2006a, b)

$$G = \frac{2\int_0^1 [u - L(u)]\,\mathrm{d}u}{1 + \rho\,|L(\theta_1)|} = \frac{1 - 2\int_0^1 L(u)\,\mathrm{d}u}{1 - \rho L(\theta_1)}, \tag{4.13}$$

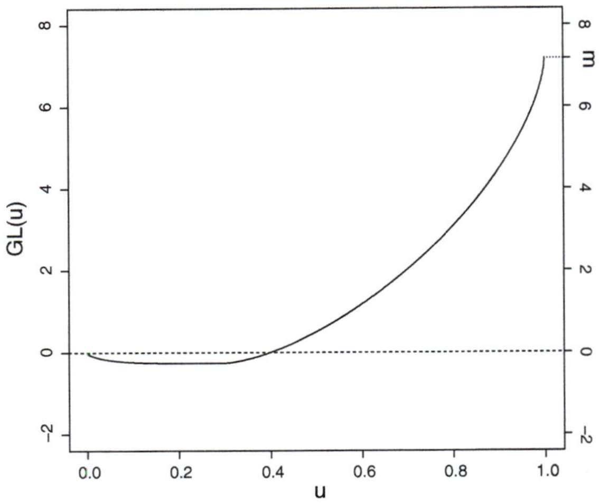

Fig. 4.2 The Generalized Lorenz curve of the mixture model for net wealth distribution

where

$$\int_0^1 L(u)\,\mathrm{d}u = \int_0^{\theta_1} L(u)\,\mathrm{d}u + \int_{\theta_1}^{\rho} L(u)\,\mathrm{d}u + \int_{\rho}^{1} L(u)\,\mathrm{d}u. \tag{4.14}$$

Using (4.11), the Gini of the κ-generalized mixture model for net wealth distribution becomes

$$G = \frac{m}{m + \rho\lambda\theta_1\Gamma\left(1+\frac{1}{s}\right)} - \frac{2\left[\frac{(1-\rho)^2\beta}{(2\kappa)^{1+\frac{1}{\alpha}}} B\left(\frac{1}{\kappa} - \frac{1}{2\alpha}, 1+\frac{1}{\alpha}\right) - \lambda\theta_1\left(1-\theta_1 2^{-1-\frac{1}{s}}\right)\Gamma\left(1+\frac{1}{s}\right)\right]}{m + \rho\lambda\theta_1\Gamma\left(1+\frac{1}{s}\right)}. \tag{4.15}$$

Hence, while β and λ are scale parameters, all the others—i.e. s, α, κ, θ_1, θ_2 and θ_3—are inequality parameters. Specifically, s, θ_1 and θ_2 account for the contribution to the Gini index of the negative and null net wealth observations, whereas α, κ and θ_3 account for the contribution of the positive observations.

The Gini coefficient of net wealth distribution is well-defined when households have negative or zero wealth holdings, but estimates of the coefficient might be greater than one if those households represent a relatively large fraction of the population; furthermore, whenever the mean net wealth is negative, estimates of the index are even negative (Amiel et al. 1996; Cowell 2011; Hagerbaumer 1977; Jenkins and Jäntti 2005; Pyatt et al. 1980). As with the Lorenz curve, the Absolute Gini index

$$AG = mG, \tag{4.16}$$

i.e. the standard Gini (4.15) multiplied by mean wealth, may be used in cases where the particular features of wealth data make standard inequality measurement tools not applicable.[3] Notice, however, that the Absolute Gini is not unit-free as the standard Gini, but it is in the units of wealth—the same holds true for the ordinates of the Generalized Lorenz curve. It becomes thus particularly important to have an appropriate price deflator and exchange rate when proceeding to compare wealth distributions across time and countries. Nevertheless, this argument does not affect the empirical results presented in the next section, which is mainly focused on the ability of the κ-generalized mixture model to account for the distinctive features of net wealth data.

4.5 Application

4.5.1 Data and Preliminary Inspection

The empirical analysis is based on data remotely accessed via the Luxembourg Wealth Study Database (LWS), a collaborative project to assemble existing micro-data on household wealth into a coherent database that aims to do for wealth what the Luxembourg Income Study Database (LIS) has achieved for income. The LWS was officially launched in 2004 and currently provides wealth data sets for various years corresponding to 12 industrialized countries—Austria, Canada, Cyprus, Finland, Germany, Italy, Japan, Luxembourg, Norway, Sweden, United Kingdom and the United States.[4] The data held in this database, drawn from national surveys and in some cases administrative sources, are harmonized as much as possible to allow for meaningful comparisons between countries, although some comparability issues still remain that are due to methodological differences across the national surveys—including differences in sampling framework, survey design and the number and definition of wealth variables recorded in each survey (for details see Sierminska et al. 2006a, b). However, despite these potential issues one must bear in mind, the LWS project remains an internationally agreed frame for the collection of household wealth data and a useful source for comparative analysis that has been used extensively in well documentable research (e.g. Cowell et al. 2012a, b; Jäntti et al. 2008, 2013, 2015; OECD 2008; Sierminska et al. 2013).

[3]The Gini coefficient is a *relative* measure of inequality, meaning that it remains invariant under equi-proportionate changes in all wealth holdings—that is, it satisfies the property of "scale invariance". By contrast, the *absolute* counterpart of the aforementioned index satisfies the property of "translation invariance", meaning that it does not alter when all the wealth values are increased or decreased by the same amount. If a relative index, when multiplied by the mean, becomes an absolute index that does not change under equal absolute translation of wealth, then it is said to possess the "compromise" property (Chakravarty 2009, Chap. 1). The Gini index is not the only one that satisfies this property.

[4]See the "LWS List of Datasets" available at http://www.lisdatacenter.org/our-data/lws-database/documentation/lws-datasets-list/.

Table 4.1 Wealth data obtained from the LWS database

Country	Year	Code[a]
Canada	1999	CA99w
Cyprus	2002	CY02w
Finland	1998	FI98w
Italy	2004	IT04w
Japan	2003	JP03w
Norway	2002	NO02w
Sweden	2002	SE02w
United Kingdom	2000	UK00w
United States	2006	USS06w

[a]Data set abbreviation to properly identify the specific country and year within the LWS database
Source "LWS List of Datasets" (http://www.lisdatacenter.org/our-data/lws-database/documentation/lws-datasets-list/)

The measure of wealth we are concerned with is total household net worth ("net worth 1" in the LWS nomenclature). This is constructed as the sum of financial and non-financial assets of the households minus total debt. As mentioned in the LWS "Variable Definition List" (http://www.lisdatacenter.org/wp-content/uploads/our-lws-documentation-variable-definition.xlsx), financial assets consist of deposit accounts, stocks, bonds and mutual funds, while non-financial assets are the sum of the values of principal residence and investment property.[5] Total household debt, in turn, refers to all outstanding loans, including home secured, vehicle loans, total installment debt, educational loans, other loans from financial institutions and informal debt.

From the LWS database we have selected all the most recent country data sets for which the "net worth 1" LWS variable is currently available for the analysis; this allows covering 9 out of 12 countries,[6] which are summarily listed in Table 4.1. All wealth data were divided by the square-root equivalence scale to adjust for differences in relative needs of households according to their size.[7] Furthermore, both the descriptive and econometric analyses use person-level adjusted weights (constructed

[5]The measure of net worth we use excludes business wealth from the definition of non-financial assets, which is otherwise included in the "net worth 2" LWS variable. Had we selected this net worth variable, however, the number of countries used in the current application would have been smaller than if business assets were excluded, since data on these assets are only available for a subset of countries (Canada, Cyprus, Finland, Italy, Norway, Sweden and the United States). We stick thus to the definition that is less inclusive but that is available for more countries.

[6]No comparable net wealth variable is indeed recorded in the LWS surveys of Austria, Germany and Luxembourg. For a chart showing which variables are available in each country data set, see the LWS "Variable Availability Matrix" (http://www.lisdatacenter.org/wp-content/uploads/LWS-data-availability-2012-03-08.xlsx).

[7]While in the case of income analysis there are good arguments for using "equivalized income" as an appropriate indicator of current individual welfare within a household, application of equivalence scales to household *wealth* data is more controversial (e.g. Sierminska and Smeeding 2005). However, since we interpret wealth as the ability to finance potential consumption, and thus as a

by multiplying the household weight by the number of household members) so that generalizations about a population can be made using results obtained from a sample of the same population.

Table 4.2 provides a number of summary statistics. Sample sizes are widely different, ranging from 349 households in Cyprus to 22,870 units in Norway. When the shares of zero and negative values are considered, Cyprus and Italy stand out for having relatively low proportions of households with nil and negative net worth, while for Norway, Sweden and the United States these proportions are far higher than in all other countries. In particular, the majority of Norwegian households (about 56 % of the population) has negative net worth and this results in some major inconsistencies, such as the fact that Norway is the only country where median net wealth is negative. However, notwithstanding these cross-country differences in the proportions of negatives and zeros, when their joint prevalence is taken into account we find it to be remarkably high on average (around 23 % of the sample size). This situation is quite different from that generally faced in the case of income data, where it is often assumed that income can only take on positive values—in practice, there may be non-positive incomes but usually the number of these is so small that one can just ignore them. By contrast, in the case of net wealth data the assumption of dealing with a positive quantity can not be justified, since it is a matter of fact that many people enter a period of indebtedness at some point in their life. Therefore, net wealth may legitimately take on negative and zero values, and the proportion of such observations could be non-negligible (as in our case) in representative samples of certain populations.

Mean net worth is also much higher than median net worth in each LWS survey. One might suspect that these large differences between the mean and the median are caused by the presence of long and heavy tails in the net wealth distributions of LWS countries, particularly at the top of the data range. Indeed, the positive skewness values listed in the eighth column of Table 4.2 suggest that the distribution of net wealth in any one country has a long tail toward the upper end, thus indicating a non-trivial prevalence of values that are "extremes" in relation to the rest of the data. Furthermore, in each of the surveys the level of kurtosis is remarkably increased above the normal value—especially in the cases of Cyprus, Norway, Sweden and the United States (ninth column of Table 4.2), meaning that the upper tails of net wealth distributions are inevitably "fat", i.e. decline to zero more slowly than exponentially. As the median would not be affected by the extreme values, this results in average net worth holdings that are consistently larger than median ones in all cases.

Additional information about the long tail in the upper end region of the net wealth distribution is provided by the sample mean excess plot shown in Fig. 4.3 for the Canadian and US data, which are likely to offer a more reliable picture of the top population percentiles due to the inclusion of a special sample for the

(Footnote 7 continued)
relevant measure of the living standard can be attained with, arguments for applying equivalence scales to adjust wealth for household size are strong (Cowell and Van Kerm 2015). The simple equivalence scale adopted here—the square root of the number of household members—is one of the most commonly used in international studies (Atkinson et al. 1995).

Table 4.2 Summary statistics on household net wealth: all LWS countries in the most recent years[a]

Code[b]	Observations	% with $W < 0$	% with $W = 0$	% with $W > 0$	Median	Mean	Skewness	Kurtosis	Gini
CA99w	15,933	18.603	2.158	79.239	27,951	66,837	7.750	115.838	0.728
CY02w	349	7.073	2.521	90.406	27,158	35,317	53.215	3,257.535	0.554
FI98w	3,893	14.232	1.346	84.422	32,015	47,973	7.568	103.718	0.659
IT04w	8,012	3.976	6.003	90.022	72,500	111,108	7.384	125.915	0.594
JP03w	2,819	11.859	6.510	81.631	7,756,718	13,900,000	4.185	42.217	0.691
NO02w	22,870	55.673	0.183	44.144	− 37,232	62,817	35.527	2,260.212	5.677
SE02w	17,953	27.261	3.135	69.604	150,345	399,052	78.910	17,812.850	0.873
UK00w	4,185	12.653	4.994	82.353	26,693	53,643	3.709	26.334	0.668
USS06w[c]	22,090	21.600	3.633	74.767	32,575	185,821	76.636	18,600.550	0.847

[a] All monetary amounts are reported in national currencies that were in use at the time of data collection
[b] See Table 4.1
[c] Due to multiple imputation procedure carried out by the data provider, data are stored as five successive replicates of each record that should not be used separately; as a result, the number of observations in the full data set is five times the actual number of respondents (see http://www.lisdatacenter.org/our-data/lws-database/documentation/lws-datasets-list/)
Source Authors' calculations based on LWS

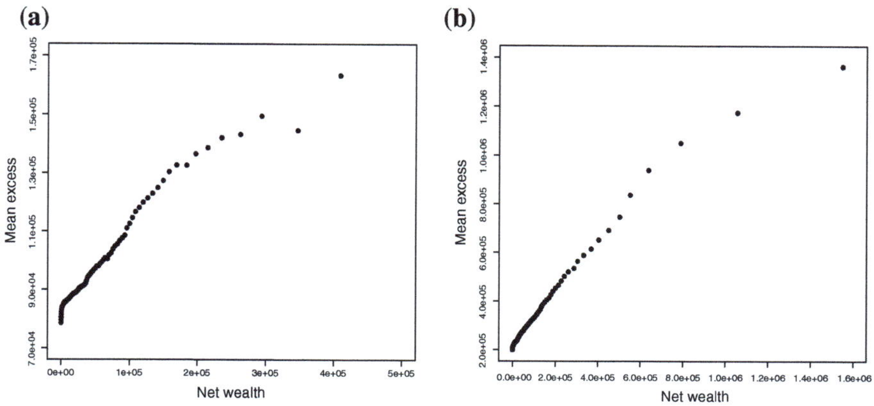

Fig. 4.3 Mean excess plots for the distributions of household net worth in Canada (**a**) and the United States (**b**). The plots have been drawn only for the positive values of net wealth corresponding to the empirical percentiles

wealthy (Jäntti et al. 2008; Sierminska et al. 2006a, b).[8] For a sequence of threshold values $(w_i)_{i=1}^{n-1}$, the mean excess plot reports the mean of exceedances over each w_i against w_i itself. Putting it differently, this is a plot of the set of ordered pairs $(w_i, e_n(w_i))$, where $e_n(w_i) = \frac{\sum_{j=i+1}^{n}(w_j - w_i)}{n-i}$ is the sample *mean excess function* and $(w_i)_{i=1}^{n-1}$ are the sample observations ranked from least to greatest. If the points in the plot show an upward trend, then this is a sign of heavy-tailed behavior. Exponentially distributed data would give an approximately horizontal line and data from a short-tailed distribution would show a downward trend. In particular, if the empirical mean excess plot seems to follow a reasonably straight line with positive slope above a certain net wealth value, then this is an indication of Pareto (power-law) behavior in tail. This is precisely the kind of behavior we observe in the LWS data for Canada and the United States. In fact, apart from some noisiness by the most extreme observations, there is evidence for consistent upward trends of the data and straightening out of the plots above some points onwards, hence providing a statistical justification for the emergence of power laws as limiting behavior for the very wealthy.

The final column of Table 4.2 shows the Gini coefficient of household net worth for each country. Sweden and the United States have the highest levels of inequality (at 0.873 and 0.847, respectively) while Cyprus and Italy have the lowest (0.554 and 0.594). Canada (0.728), Finland (0.659), Japan (0.691) and the United Kingdom (0.668) are positioned in the middle of these extremes. The estimate of the Gini coefficient for Norway takes on a value that is considerably greater than one (5.677). The explanation rests on the very high proportion of Norwegian households with

[8] Properties of the mean excess plot are reviewed, for instance, in Beirlant et al. (2004). Since we are interested here in the upper tail behavior of the distribution, the plots have been drawn only for the positive values of net wealth corresponding to the empirical percentiles.

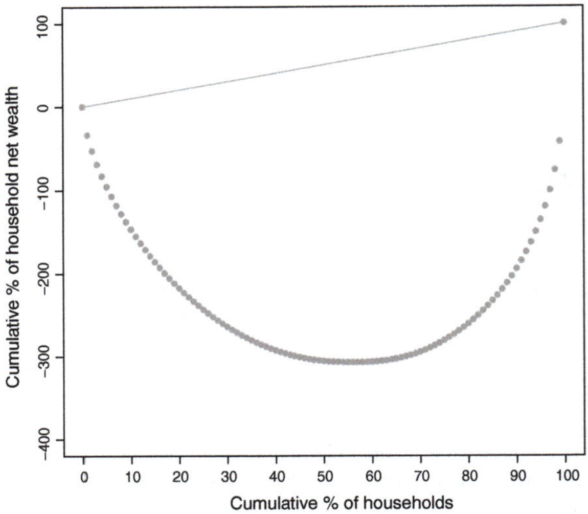

Fig. 4.4 Lorenz curve for net wealth in Norway. The data points correspond to the percentiles of the distribution

negative net worth: the Lorenz curve lies completely below the horizontal axis in this case (see Fig. 4.4) and so twice the area between the curve and the ray of perfect equality (equal to the Gini) is substantially greater than one—see the discussion of the previous section. But perhaps the greatest surprise is Sweden, which despite being one of the most equal countries in terms of income distribution, it is ranked as one of the most unequal countries in terms of wealth—even more so than the United States. Thus, there is no reason to expect that there would be lower wealth inequality in countries characterized by lower income inequality, as the wealth inequality ranking of countries may be different at times from the respective ranking in terms of income inequality (Brandolini 2006; Jäntti et al. 2008).

4.5.2 Estimation and Comparison of Finite Mixture Models for Net Wealth Distribution

Table B.4 presents the parameter estimates arising from the fit of the κ-generalized mixture model (4.7) to the LWS data. For the ease of comparison, the results of fitting alternative net wealth distribution models are also shown. Namely, these models are

$$F(w) = \begin{cases} \theta_1 \exp\left[-\left(\frac{|w|}{\lambda}\right)^s\right] & \text{if } w < 0, \\ \rho & \text{if } w = 0, \\ \rho + (1-\rho)\left\{1 - \left[1 + \left(\frac{w}{b}\right)^a\right]^{-q}\right\} & \text{if } w > 0, \end{cases} \tag{4.17}$$

where the κ-generalized specification is replaced by the Singh-Maddala distribution for the positive values of net wealth (Jäntti et al. 2015), and

$$F(w) = \begin{cases} \theta_1 \exp\left[-\left(\frac{|w|}{\lambda}\right)^s\right] & \text{if } w < 0, \\ \rho & \text{if } w = 0, \\ \rho + (1-\rho)\left\{\left[1 + \left(\frac{w}{b}\right)^{-a}\right]^{-p}\right\} & \text{if } w > 0, \end{cases} \tag{4.18}$$

where the Dagum type I distribution is assumed instead of the κ-generalized (Cowell and Van Kerm 2015; Jenkins and Jäntti 2005). The parameters were estimated in all cases by minimizing the negative of the log-likelihood function via a modified Newton-Raphson procedure implemented in Stata's `ml` command (Gould et al. 2010), with parameter standard error estimates based on the negative inverse Hessian. Convergence was achieved easily within several iterations.[9]

The small value of the errors indicates that all the parameters were very precisely estimated. The mixture proportions (the θ's) correspond exactly to the sample estimates shown in Table 4.2, and the scale parameters (λ, the b's and β) vary in line with differences in the scale of the net wealth distribution across countries. The other parameters (s, the a's, α, q, p and κ) reflect differences in distributional shape, and are easiest to interpret by comparing summary statistics derived from the model parameters to the statistics computed from the raw data. For instance, judging by the absolute percentage error (3.64) and excepting for Norway, Fig. 4.5 shows that the mean net wealth and Absolute Gini coefficient as estimated from the three mixture models are very close to their sample estimates—to facilitate such a comparison, the predicted values and the actual sample estimates for the distributional statistics considered are given in Table 4.3.[10] In particular, the agreement between the

[9]This was not the case for the UK household net worth distribution (UK00w), where the maximum likelihood estimation procedure for the Singh-Maddala mixture model failed to converge to a solution even after fiddling with different starting guess values and optimization routines. Hence, the corresponding results have not been reported in Table B.4.

[10]Mean net wealth estimates for the κ-generalized mixture model have been obtained by substituting the estimated parameters into Eq. (4.10). Mean predictions from the Singh-Maddala and Dagum type I models, in turn, were derived by replacing $E_3(W)$ in formula (4.10) with $E_3(W) = \frac{b\Gamma\left(1+\frac{1}{a}\right)\Gamma\left(q-\frac{1}{a}\right)}{\Gamma(q)}$ and $E_3(W) = \frac{b\Gamma\left(p+\frac{1}{a}\right)\Gamma\left(1-\frac{1}{a}\right)}{\Gamma(p)}$ for, respectively, the Singh-Maddala and Dagum type I specifications (see e.g. Clementi et al. 2012b). As for the Gini, we stick here to the absolute definition (4.16) of the index because of the high proportion of negative net wealth holders in Norway that causes a major inconsistency in the computation of the conventional Gini for this country (see Sect. 4.5.1). The standard Gini definition for the κ-generalized model of net wealth distribution is (4.15), whereas for the Singh-Maddala and Dagum type I mixtures, respectively, it is given by (Clementi et al. 2012b)

$$G = \frac{m - 2\left[(1-\rho)^2 bqB\left(2q - \frac{1}{a}, 1 + \frac{1}{a}\right) - \lambda\theta_1\left(1 - \theta_1 2^{-1-\frac{1}{s}}\right)\Gamma\left(1 + \frac{1}{s}\right)\right]}{m + \rho\lambda\theta_1\Gamma\left(1 + \frac{1}{s}\right)}$$

and

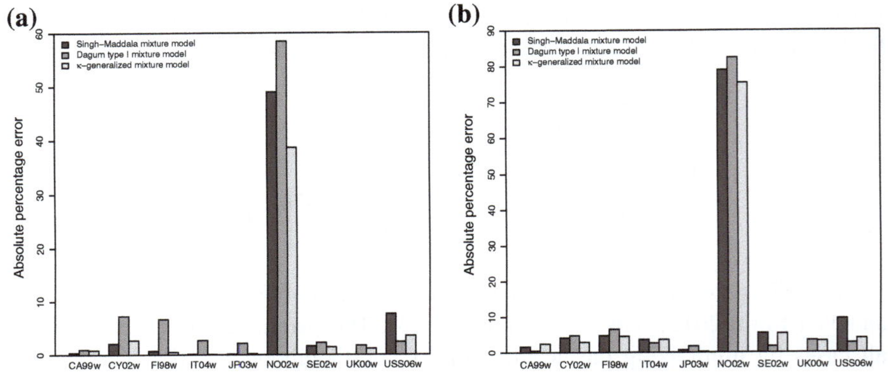

Fig. 4.5 Absolute percentage error between the implied and sample estimates of the mean (**a**) and Absolute Gini coefficient (**b**): all LWS countries in the most recent years. Due to non-convergence of the maximum likelihood algorithm, the errors corresponding to the Singh-Maddala mixture model are not shown for the UK household net worth distribution

implied and sample estimates of the mean and Absolute Gini is much closer for the κ-generalized mixture model in five cases out of nine,[11] thus indicating a slightly better performance of this model than its competitors.

The results summarized in Table B.4 allow us to emphasize a distinctive feature of wealth distributions, i.e. the concentration of density mass at zero. There is often a marked spike at zero because a relatively large fraction of the population has no wealth. Similar spikes do not occur with income distributions, where it is often the case that the density mass vanishes when income goes to zero. As the Weibull, Singh-Maddala, Dagum type I and κ-generalized distributions are zero-modal with a pole at the origin if, respectively, $s < 1, a \leq 1, ap \leq 1$ and $\alpha \leq 1$, it is often verified from the estimates of these parameters that the probability density functions of the three mixture models inevitably transfer some density mass from the neighboring values to the cited spike at zero—i.e. they diverge, rather than vanish, when the argument goes to zero from both the negative and positive ends of the wealth range.[12] This

(Footnote 10 continued)

$$G = \frac{m - 2\left\{(1-\rho)^2 bp\left[B\left(p+\frac{1}{a}, 1-\frac{1}{a}\right) - B\left(2p+\frac{1}{a}, 1-\frac{1}{a}\right)\right]\right\}}{m + \rho\lambda\theta_1 \Gamma\left(1+\frac{1}{s}\right)} + \frac{2\lambda\theta_1\left(1-\theta_1 2^{-1-\frac{1}{s}}\right)\Gamma\left(1+\frac{1}{s}\right)}{m + \rho\lambda\theta_1\Gamma\left(1+\frac{1}{s}\right)}.$$

[11]Corresponding all at once to the net worth distributions of Cyprus (CY02w), Finland (FI98w), Italy (IT04w), Japan (JO03w), Norway (NO02w), Sweden (SE02w), the United Kingdom (UK00w) and including the case of Norway (NO02w), where particularly large errors are found for both the mean and Absolute Gini predictions by all three models.

[12]The behavior around the mode of Weibull, Singh-Maddala and Dagum type I distributions is reviewed, e.g., in Kleiber and Kotz (2003). For the κ-generalized distribution see Chap. 3.

Table 4.3 Selected distributional statistics implied by the Singh-Maddala, Dagum type I and κ-generalized mixture models of net wealth distribution: all LWS countries in the most recent years

Code[a]	Mean[b]				Absolute Gini[b]			
	Observed	SM	D	κ-gen	Observed	SM	D	κ-gen
CA99w	66,837	67,106	66,192	67,400	48,635	49,447	48,421	49,842
CY02w	35,317	34,571	32,764	34,360	19,573	20,384	18,643	20,140
FI98w	47,973	48,316	44,770	48,200	31,603	33,110	29,563	33,014
IT04w	111,108	111,269	108,126	111,225	65,973	68,356	64,228	68,340
JP03w	13,900,000	13,876,147	14,193,797	13,924,628	9,605,596	9,544,109	9,771,212	9,587,117
NO02w	62,817	32,065	26,088	38,497	356,620	75,229	62,797	88,056
SE02w	399,052	392,483	408,074	393,283	348,318	329,461	342,335	329,814
UK00w[c]	53,643	–	54,597	54,270	35,828	–	37,056	36,991
USS06w	185,821	171,554	190,299	192,429	157,316	142,339	161,604	163,518

[a] See Table 4.1
[b] *SM* = Singh-Maddala mixture model; *D* = Dagum type I mixture model; κ-gen = κ-generalized mixture model
[c] The statistics implied by the Singh-Maddala mixture model are not shown due to non-convergence of the maximum likelihood algorithm
Source Authors' calculations based on LWS

finding of a divergent probability density in the limit of zero wealth is also shared by other studies on the distribution of wealth (e.g. Drăgulescu and Yakovenko 2001).

The parameter estimates reported in Table B.4 were also employed to build estimated Lorenz curves by applying Eq. (4.11) for the κ-generalized mixture model and using

$$L(u) = \begin{cases} -\frac{\lambda\theta_1}{m}\Gamma\left(1+\frac{1}{s}, \log\frac{\theta_1}{u}\right), & \text{if } 0 \le u < \theta_1, \\ -\frac{\lambda\theta_1}{m}\Gamma\left(1+\frac{1}{s}\right), & \text{if } \theta_1 \le u \le \rho, \\ \frac{1}{m}\left\{(1-\rho)\, bq\left[B\left(q-\frac{1}{a}, 1+\frac{1}{a}\right)\right.\right. & \\ \left.-B\left(\left[\frac{1-u}{1-\rho}\right]^{\frac{1}{q}}; q-\frac{1}{a}, 1+\frac{1}{a}\right)\right] & \text{if } u > \rho, \\ \left.-\lambda\theta_1\Gamma\left(1+\frac{1}{s}\right)\right\} & \end{cases} \tag{4.19}$$

and

$$L(u) = \begin{cases} -\frac{\lambda\theta_1}{m}\Gamma\left(1+\frac{1}{s}, \log\frac{\theta_1}{u}\right), & \text{if } 0 \le u < \theta_1, \\ -\frac{\lambda\theta_1}{m}\Gamma\left(1+\frac{1}{s}\right), & \text{if } \theta_1 \le u \le \rho, \\ \frac{1}{m}\left\{(1-\rho)\, bpB\left(\left[\frac{u-\rho}{1-\rho}\right]^{\frac{1}{p}}; p+\frac{1}{a}, 1\right.\right. & \text{if } u > \rho \\ \left.\left.-\frac{1}{a}\right)-\lambda\theta_1\Gamma\left(1+\frac{1}{s}\right)\right\} & \end{cases} \tag{4.20}$$

for, respectively, the Singh-Maddala and Dagum type I net wealth distribution models.[13] The curves, together with the empirical estimates, are presented in Fig. 4.6 for three LWS countries with varying institutional and welfare regimes—as well as different levels of wealth inequality: Italy, Japan and the United States. Even if it is small, one can see a difference between the three predictions, in that the Lorenz curve for US net wealth estimated from the Singh-Maddala mixture model lies slightly above the empirical one for approximately the top 20 % of the wealthiest households, while the Dagum type I and κ-generalized specifications lead to estimated Lorenz curves exhibiting a degree of inequality that is much more in line with the observed one—see Fig. 4.6c. Furthermore, we see in Fig. 4.6a that the Singh-Maddala and κ-generalized Lorenz curves lie below the Italian curve from a population share of around 50 % and up until 90 %, whereas the curve associated with the Dagum type I mixture model fits the data very well in the whole range. Finally, Fig. 4.6b shows that the Lorenz curves as estimated from the three mixture models are all very close to their sample counterpart for Japanese net worth data.

[13]The formulas of the Lorenz curve for the Singh-Maddala and Dagum type I mixture models were derived by Clementi et al. (2012b).

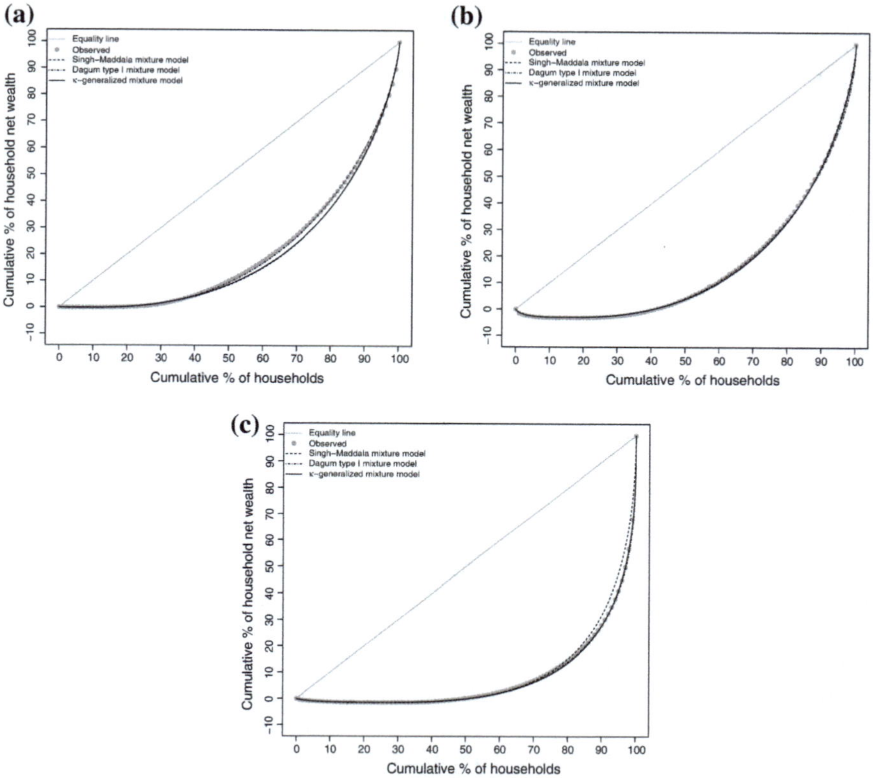

Fig. 4.6 Observed and predicted Lorenz curves for net wealth in Italy (**a**), Japan (**b**) and the United States (**c**). The data points correspond to the empirical percentiles of the distributions

The above evidence holds vis-à-vis a further check involving goodness-of-fit indicators such as the root mean squared error, defined as the square root of the average squared error between the observed and predicted values of the Lorenz curve. In mathematical terms, this is expressed as

$$LRMSE = \sqrt{\frac{1}{n}\sum_{i=1}^{n}\left[L_i - L\left(u_i;\hat{\boldsymbol{\theta}}\right)\right]^2}, \tag{4.21}$$

where L_i denotes the ith empirical Lorenz ordinate and $L\left(u_i;\hat{\boldsymbol{\theta}}\right)$ is the corresponding theoretical ordinate deduced from the fitted mixture models with parameter estimates $\hat{\boldsymbol{\theta}}$. Clearly, lower values of $LRMSE$ indicate a better fit. The comparison results between the competing models based on the above criterion are shown in Table 4.4. As can be seen, the κ-generalized model of net wealth ranks first for all

Table 4.4 Goodness-of-fit comparisons for estimated Lorenz curves of household net wealth in Italy, Japan and the United States

Code[a]	Model[b]	$LRMSE \times 10^2$	Rank
IT04w	SM	0.1833	3
	D	0.0595	1
	κ-gen	0.1767	2
JP03w	SM	0.0572	3
	D	0.0546	2
	κ-gen	0.0440	1
USS06w	SM	0.3433	3
	D	0.2528	2
	κ-gen	0.1620	1

[a]See Table 4.1
[b]*SM* = Singh-Maddala mixture model; *D* = Dagum type I mixture model; κ-gen = κ-generalized mixture model
Source Authors' calculations based on LWS

countries investigated but Italy, where it is outperformed by the Dagum type I mixture model. Similar results holds for the Generalized Lorenz curves derived from Eq. (4.12), a graphical display of which is given in Fig. 4.7.

Finally, Fig. 4.8 presents the relationship between the log-rank and log-size of positive net wealth for Sweden, the LWS country exhibiting the highest level of wealth inequality. This double-logarithmic framework, known as the Zipf plot, has been used rarely in economics, but is more common in physics (see e.g. Takayasu 1990).[14] In particular, it is natural to use when focusing on the top part of the distribution because it accentuates the upper tail, making it easier to detect deviations in that part of the distribution from the theoretical prediction of a particular model. The lines shows the predicted Zipf plots obtained from the fit of the models considered. As the figure reveals, all of them are in good agreement with the actual data in the positive range of the net wealth distribution. However, at the top tail the theoretical Zipf plot for the κ-generalized mixture model lies somewhat much closer to the empirical one than in the cases of the Singh-Maddala and Dagum type I specifications. The impression left by the graphical display is confirmed by calculation of the Anderson-Darling distance between the empirical distribution function and the hypothesized models, which results in a lower value of the statistic (0.035) for the κ-generalized model as compared to the Singh-Maddala (0.043) and Dagum type I (0.039) mixture models.[15]

[14]Let $(w_i)_{i=1}^n$ be a set of n wealth holdings for which the cumulative distribution function is $\hat{F}_i = \frac{i}{n}$ and suppose that the observations are ordered from largest to smallest, so that the index i is the rank of w_i. The Zipf plot of the sample is the graph of $\ln(i)$ against $\ln(w_i)$. Because of the ranking, $\frac{i}{n} = 1 - \hat{F}_i$, so $\ln(i) = \ln\left[1 - \hat{F}_i\right] + \ln(n)$. Thus, the log of the rank is simply a transformation of the cumulative distribution function. For an illustration of basic properties of the Zipf plot see for instance Stanley et al. (1995).

[15]The reported values of the Anderson-Darling statistic are multiplied by 100. This statistic is known to be more powerful than other measures quantifying the distance between the empirical

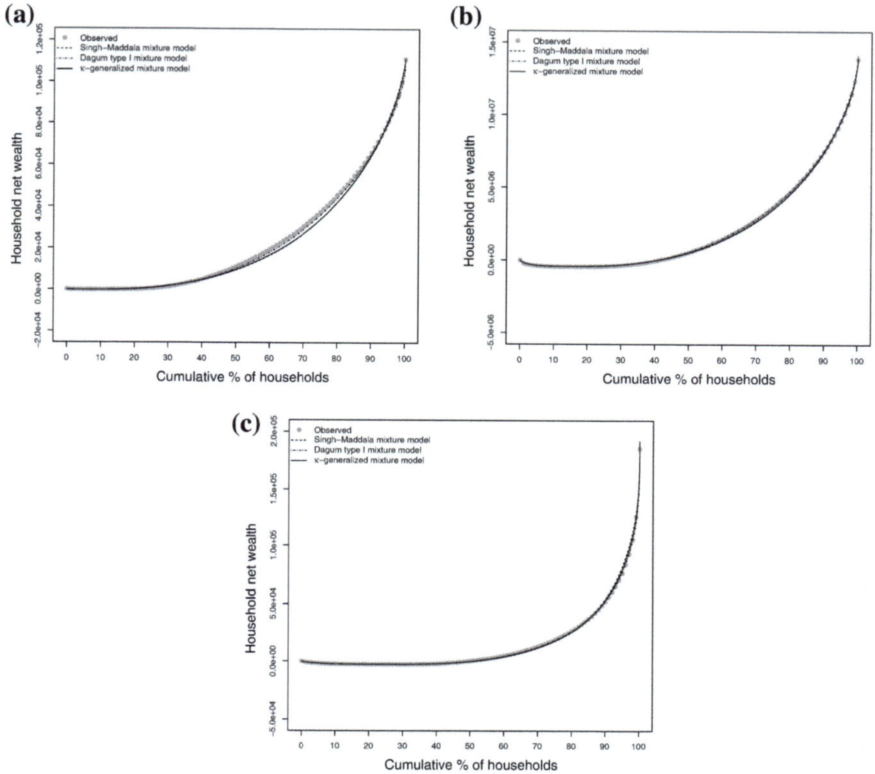

Fig. 4.7 Observed and predicted Generalized Lorenz curves for net wealth in Italy, Japan and the United States. The data points correspond to the empirical percentiles of the distributions. **a** Italy. **b** Japan. **c** United States

This point is of particular relevance in the current context, both for the documented presence of long and fat tails towards the upper end of the net wealth distribution and the fact that all of the three densities accounting for the positive range of wealth obey the weak Pareto law (Mandelbrot 1960). The weak version of the Pareto law states that the right-hand tail of a distribution behaves in the limit as a simple Pareto model, with anexponent that is a function of the parameters governing the shape of

(Footnote 15 continued)
distribution function of a uni-variate data set and the cumulative distribution function of a reference distribution, since it provides equal sensitivity at the tails as at the median of the distribution (e.g. Thode 2002). The formula used for the Anderson-Darling distance is the one reported by Monahan (2011, p. 358), which allows for weighted observations.

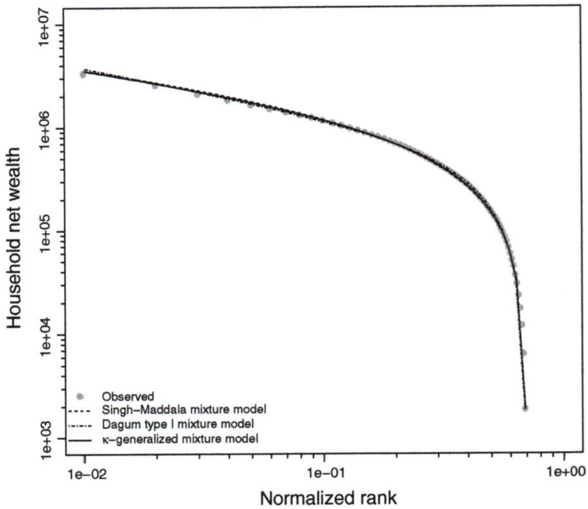

Fig. 4.8 Zipf plot (double-logarithmic plot of wealth vs. rank) for the positive values of household net wealth in Sweden. The data points correspond to the empirical percentiles of the distribution while the lines are the predicted Zipf plots obtained from the fit of the models under scrutiny

the distribution (see e.g. Krämer and Ziebach 2004, for an overview).[16] Hence, finite mixture models should deserve the attention of wealth analysts as they allow them to account for the well-documented Pareto functional form at the high-wealth range without disregarding the fact that what is distinctive about wealth distributions in contrast to income distributions is the presence of negative and zero values that can be neglected only if one is willing to bear the cost of omitting a significant part of the story to be told.

[16]The values of the Paretian upper tail index can be derived from parameter estimates of the Singh-Maddala, Dagum type I and κ-generalized distributions, respectively, as $\gamma = aq$, $\gamma = ap$ and $\gamma = \frac{\alpha}{\kappa}$. For more details on the upper tail behavior of the κ-generalized distribution see Sect. 3.1.2. For the Singh-Maddala and Dagum type I distributions see instead Wilfling and Krämer (1993) and Kleiber (1996, 2008a).

Chapter 5
Four-Parameter Extensions of the κ-Generalized Distribution

Concern for man and his fate must always form the chief interest of all technical endeavours. Never forget this in the midst of your diagrams and equations.

Albert Einstein

Abstract This chapter presents recent four-parameter extensions of the κ-generalized distribution and surveys their applications to households income/consumption data of different countries and years as well as empirical comparisons with other existing four-parameter income distribution models.

Keywords Income and consumption distributions · Parametric modelling · Four-parameter κ-generalized distributions

5.1 Definitions and Basic Properties

Two four-parameter extensions of the κ-generalized distribution were proposed by Okamoto (2013) and named *extended κ-generalized distributions of the first and second kind* (or EκG1 and EκG2, for short).

The first kind of generalization, the EκG1, was obtained by "Weibullizing" a two-parameter deformed exponential, implicitly defined as the inverse of the following two-parameter generalization of the logarithm function (Kaniadakis et al. 2004, 2005)

$$\ln_{\{\kappa,r\}}(x) = x^r \frac{x^\kappa - x^{-\kappa}}{2\kappa}, \tag{5.1}$$

which recovers the standard logarithm in the limit $(\kappa, r) \to (0, 0)$ independently on the direction.

The content of this chapter is adapted from Okamoto (2013).

F. Clementi and M. Gallegati, *The Distribution of Income and Wealth*,
New Economic Windows, DOI 10.1007/978-3-319-27410-2_5

No closed-form expression of the cumulative distribution function is available for the EκG1, while the quantile function is given by the pleasantly simple formula

$$F^{-1}(u; a, b, q, r) = b\left[-(1-u)^r \frac{(1-u)^{\frac{1}{2q}} - (1-u)^{-\frac{1}{2q}}}{1/q}\right]^{\frac{1}{a}}, \quad (5.2)$$

with $0 < u < 1$, where $a, b, q > 0$ and $r < \frac{1}{2q}$. When $r = 0$ (and $a = \alpha$, $b = \beta$, $q = \frac{1}{2\kappa}$) the EκG1 quantile function is equivalent to that of the κ-generalized distribution—compare with (3.21). When $r = -\frac{1}{2q}$, the EκG1 is also similar to the Singh-Maddala distribution with density function given by (3.62).

The probability density function of the EκG1 can be expressed in terms of the cumulative probabilities u as follows[1]

$$f_u(u; a, b, q, r) = \frac{a\left[-(1-u)^r \frac{(1-u)^{\frac{1}{2q}} - (1-u)^{-\frac{1}{2q}}}{1/q}\right]^{-\frac{1}{a}+1}}{b\left[\left(qr + \frac{1}{2}\right)(1-u)^{r+\frac{1}{2q}-1} - \left(qr - \frac{1}{2}\right)(1-u)^{r-\frac{1}{2q}-1}\right]}. \quad (5.3)$$

Since $x = F^{-1}(u; a, b, q, r)$ and $u = F(x; a, b, q, r)$ for any pair of values (x, u), it follows from the definition of differentiation that

$$f_u(u; a, b, q, r) = \frac{1}{\frac{\mathrm{d}F^{-1}(u;a,b,q,r)}{\mathrm{d}u}} = \frac{\mathrm{d}F(x; a, b, q, r)}{\mathrm{d}x} = f(x; a, b, q, r). \quad (5.4)$$

This result provides a way of plotting the density function (5.3). Indeed, if we let u take the values, say, $u = 0.01, 0.02, \ldots, 0.99$, and plot the points $\left(F^{-1}(u; a, b, q, r), f_u(u; a, b, q, r)\right)$, then we get the plot of points $(x, f(x; a, b, q, r))$, i.e. the plot of the density of x given by (5.4). Thus plots of the EκG1 probability density function can be obtained entirely from the quantile function (5.2).

The EκG1 density follows a power-law behavior in both of its tails, i.e.

$$f(x; a, b, q, r) \underset{x\to 0}{\sim} c_1 x^{a-1} \quad \text{and} \quad f(x; a, b, q, r) \underset{x\to +\infty}{\sim} c_2 x^{-\frac{a}{\left(\frac{1}{2q}-r\right)}-1}, \quad (5.5)$$

where c_1 and c_2 are positive constants. The upper-tail power-law behavior is simply the Pareto law, whereas the power-law behavior in the lower tail of the distribution is a stylized fact that sometimes seems to be born out in actual data (e.g. Reed 2003, 2004) and was indeed identified many years ago by Champernowne (1953).

Figure 5.1 illustrates the shape of the EκG1 density for various parameter values. In the case $a > 1$, as in panels (a) and (b), $f(x; a, b, q, r) \to 0$ when $x \to 0$ and there is always an interior mode. By contrast, if $a = 1$, $f(x; a, b, q, r) \to \frac{a}{b}$ when

[1] A probability density function expressed in terms of u was called the "density quantile function" by Parzen (1979).

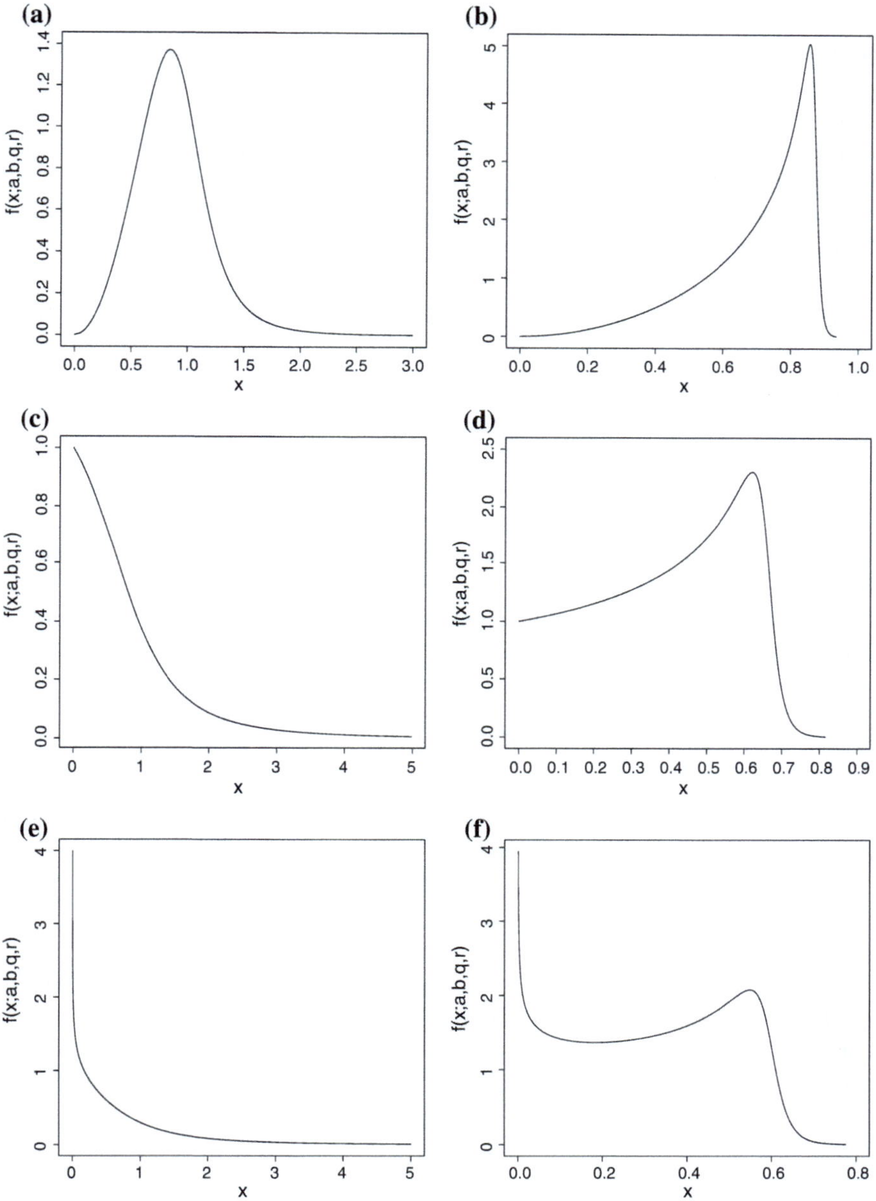

Fig. 5.1 EκG1 density for various parameter values: **a** $a = 3.00, b = 1.00, q = 0.60$ and $r = 0.30$; **b** $a = 3.00, b = 1.00, q = 0.60$ and $r = 0.80$; **c** $a = 1.00, b = 1.00, q = 0.60$ and $r = 0.30$; **d** $a = 1.00, b = 1.00, q = 0.60$ and $r = 0.80$; **e** $a = 0.80, b = 1.00, q = 0.60$ and $r = 0.30$; **f** $a = 0.80, b = 1.00, q = 0.60$ and $r = 0.80$

$x \to 0$, i.e. the EκG1 density approaches a finite positive value at the left limit; furthermore, the density is monotonically decreasing for $r \leq \frac{1}{2}$, whereas for $r > \frac{1}{2}$ it permits an interior mode—see panels (c) and (d), respectively. Finally, for $a < 1$ the EκG1 density is infinite at the left limit, i.e. $f(x; a, b, q, r) \to +\infty$ when $x \to 0$, and it can be either monotonically decreasing—as in panel (e)—or have a local maximum and minimum—as in panel (f).[2]

The second generalization, the EκG2, was derived by extending the κ-generalized Lorenz curve

$$L(u) = I_z\left(1 + \frac{1}{\alpha}, \frac{1}{2\kappa} - \frac{1}{2\alpha}\right), \quad z = 1 - (1-u)^{2\kappa} = I_u^{-1}\left(1, \frac{1}{2\kappa}\right), \tag{5.6}$$

in the following manner

$$L(u) = I_z\left(p + \frac{1}{a}, q - \frac{1}{2a}\right), \quad z = I_u^{-1}(p, q), \tag{5.7}$$

where $a = \alpha, q = \frac{1}{2\kappa}$ and $I_u^{-1}(\cdot, \cdot)$ denotes the inverse of the regularized incomplete beta function.

The cumulative distribution function of the EκG2 is available in closed form. It can be expressed in terms of the incomplete beta function ratio as follows

$$F(x; a, b, p, q) = I_z(p, q), \quad z = \left(\frac{x}{b}\right)^a \left[\sqrt{1 + \frac{1}{4}\left(\frac{x}{b}\right)^{2a}} - \frac{1}{2}\left(\frac{x}{b}\right)^a\right], \tag{5.8}$$

where $x > 0$ and all four parameters a, b, p, q are positive. Equation (5.8) implies that even the quantile function is available in closed form; it is

$$F^{-1}(x; a, b, p, q) = bz^{\frac{1}{a}}(1-z)^{-\frac{1}{2a}}, \quad z = I_u^{-1}(p, q). \tag{5.9}$$

For $p = 1$ (and $a = \alpha, b = \beta, q = \frac{1}{2\kappa}$), it follows from (5.8) and (5.9) that the EκG2 is equivalent to the κ-generalized distribution. Furthermore, the EκG2 can be considered as a new kind of generalized beta distribution, being related to the GB1 and GB2 proposed by McDonald (1984) and his associates via the relations

$$Y \sim \text{GB2}(1, 1, p, q) \Longrightarrow X = \frac{Y}{\sqrt{1+Y}} \sim \text{E}\kappa\text{G2}(1, 1, p, q) \tag{5.10}$$

[2]The existence of the maximum and minimum points is guaranteed by satisfaction of the following three inequalities

$$A < 0, \quad 0 < B < -2A \quad \text{and} \quad B^2 > 4AC,$$

with $A = \frac{2q(a-r)-1}{a}\left(r + \frac{1}{2q}\right)$, $B = -A - C + \frac{2(a-1)}{aq}$ and $C = \frac{2q(a-r)+1}{a}\left(r - \frac{1}{2q}\right)$.

and

$$Z = \frac{Y}{1+Y} \sim \text{GB1}\,(1, 1, p, q) \Longrightarrow X = \sqrt{YZ} \sim \text{E}\kappa\text{G2}\,(1, 1, p, q)\,. \qquad (5.11)$$

The probability density function of the EκG2 has the form

$$f\,(x; a, b, p, q) = \frac{a}{bB\,(p, q)} \frac{z^{p-\frac{1}{a}}\,(1-z)^{q+\frac{1}{2a}}}{1-\frac{1}{2}z}, \qquad (5.12)$$

where $z = \left(\frac{x}{b}\right)^a \left[\sqrt{1+\frac{1}{4}\left(\frac{x}{b}\right)^{2a}} - \frac{1}{2}\left(\frac{x}{b}\right)^a\right]$. The lower and upper tails of (5.12) exhibit power-law (Paretian) behavior, i.e.

$$f\,(x; a, b, p, q) \underset{x\to 0}{\sim} c_3 x^{ap-1} \quad \text{and} \quad f\,(x; a, b, p, q) \underset{x\to +\infty}{\sim} c_4 x^{-2aq-1} \qquad (5.13)$$

for constants c_3 and c_4.

Figure 5.2 charts the EκG2 density for various parameter values. In the case $p > \frac{1}{a}$, as in panels (a) and (b), $f\,(x; a, b, p, q) \to 0$ when $x \to 0$ and the density always exhibits an interior mode. By contrast, if $p = \frac{1}{a}$, $f\,(x; a, b, p, q) \to \frac{a}{bB(p,q)}$ when $x \to 0$, i.e. the EκG2 density approaches a finite positive value at the left limit; furthermore, the density is monotonically decreasing for $q + \frac{1}{2a} \geq \frac{1}{2}$, whereas for $q + \frac{1}{2a} < \frac{1}{2}$ it permits an interior mode—see panels (c) and (d), respectively. Finally, for $p < \frac{1}{a}$ the EκG2 density is infinite at the left limit, i.e. $f\,(x; a, b, p, q) \to +\infty$ when $x \to 0$, and it can be either monotonically decreasing—as in panel (e)—or have a local maximum and minimum—as in panel (f).[3]

Venter (1983) observed that the reciprocal of a random variable from the GB2 distribution with parameters a, b, p and q still follows a GB2 distribution with parameters a, $\frac{1}{b}$, p and q, i.e.

$$X \sim \text{GB2}\,(a, b, p, q) \Longleftrightarrow \frac{1}{X} \sim \text{GB2}\left(a, \frac{1}{b}, p, q\right). \qquad (5.14)$$

Although somewhat related, this closure property under inversion is special to the GB2 and does not extend to the four-parameter variants of the κ-generalized distribution, in the sense that the reciprocal of a random variable from the EκG1 or EκG2 distribution does not follow the EκG1/EκG2. Rather, the reciprocal of a random variable from the EκG1 distribution with parameters a, b, q and r follows the *inverse of the extended κ-generalized distribution of the first kind* (for short referred

[3]The conditions to be met for the existence of the maximum and minimum points are as follows

$$(p+2q)^2 + 2\left(p - 2q - \frac{2}{a}\right) + 1 > 0, \quad p + q - \frac{1}{2a} - 1 < 0 \quad \text{and} \quad 3p + 2q - \frac{2}{a} - 1 < 0.$$

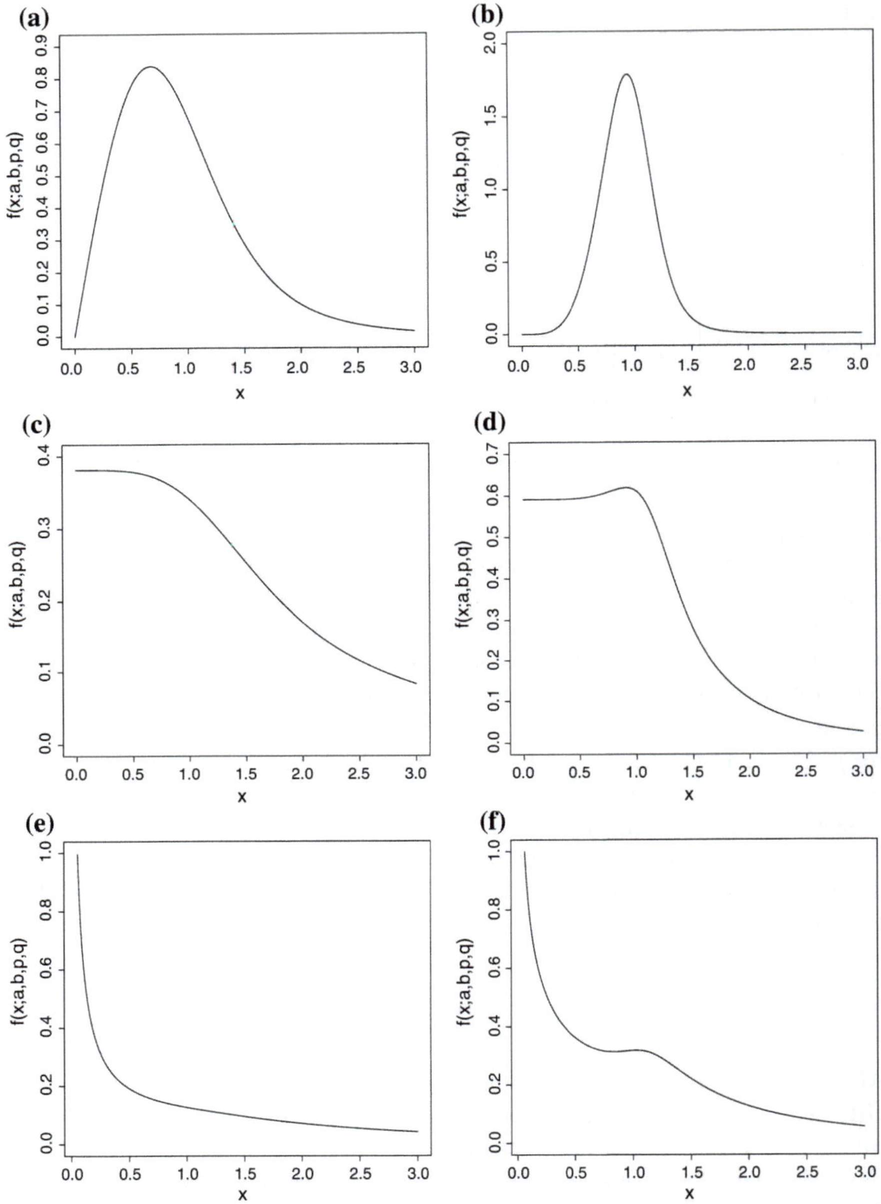

Fig. 5.2 EκG2 density for various parameter values: **a** $a = 2.00$, $b = 1.00$, $p = 1.00$ and $q = 1.00$; **b** $a = 5.00$, $b = 1.00$, $p = 1.00$ and $q = 1.00$; **c** $a = 2.00$, $b = 1.00$, $p = 0.50$ and $q = 0.25$; **d** $a = 5.00$, $b = 1.00$, $p = 0.20$ and $q = 0.25$; **e** $a = 2.00$, $b = 1.00$, $p = 0.10$ and $q = 0.10$; **f** $a = 5.00$, $b = 1.00$, $p = 0.10$ and $q = 0.10$

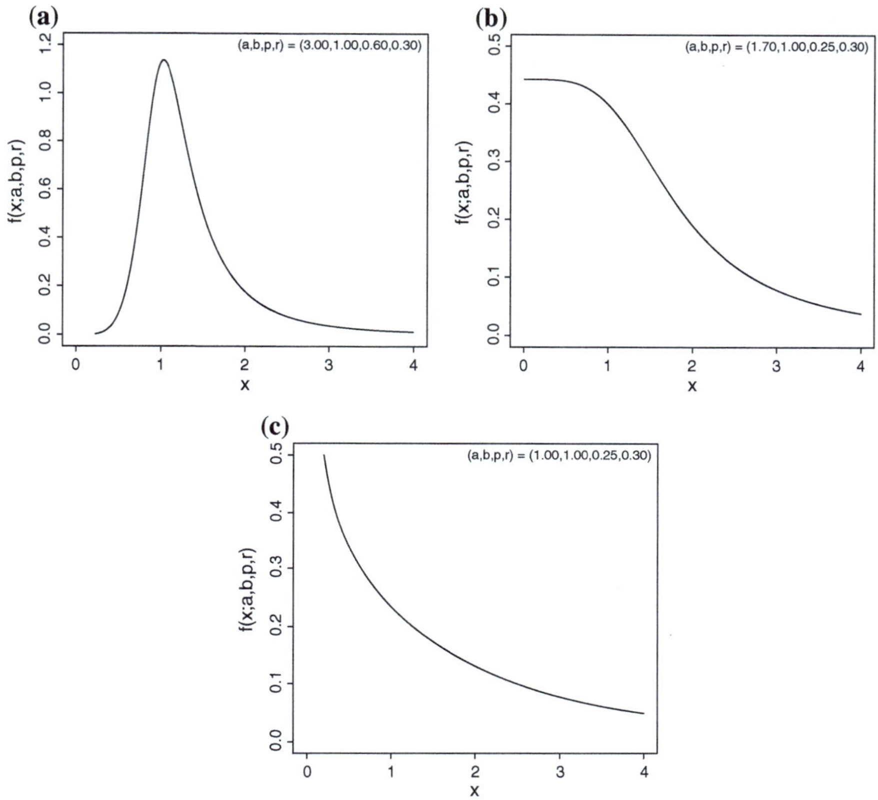

Fig. 5.3 IEκG1 density for various parameter values: **a** $a > \frac{1}{2p} - r$; **b** $a = \frac{1}{2p} - r$; **c** $a < \frac{1}{2p} - r$

to as IEκG1) with parameters a, b, p and r. Similarly, the inverse of the EκG2 with parameters a, b, p and q is the IEκG2 distribution with parameters a, b, p and q. The density, distribution function and quantile function of these two inverses are summarily listed in Table B.5. Note that the parameters of the IEκG1 must be in a domain defined as $a, b, p > 0$ and $r < \frac{1}{2p}$, whereas all four parameters a, b, p and q of the IEκG2 must be positive.

The IEκG1 also has power-law tails, that is

$$f(x; a, b, p, r) \underset{x \to 0}{\sim} c_5 x^{\frac{a}{\left(\frac{1}{2p} - r\right)^{-1}}} \quad \text{and} \quad f(x; a, b, p, r) \underset{x \to +\infty}{\sim} c_6 x^{-a-1}. \qquad (5.15)$$

When $x \to 0$, the IEκG1 density approaches zero or a finite positive value or diverges to infinity depending on whether, respectively, $a > \frac{1}{2p} - r$, $a = \frac{1}{2p} - r$ or $a < \frac{1}{2p} - r$. In the case $a > \frac{1}{2p} - r$, the IEκG1 density has an interior mode; otherwise, it is always monotonically decreasing without further restrictions on the parameters—unlike the EκG1 density. Figure 5.3 provides some examples of IEκG1 densities.

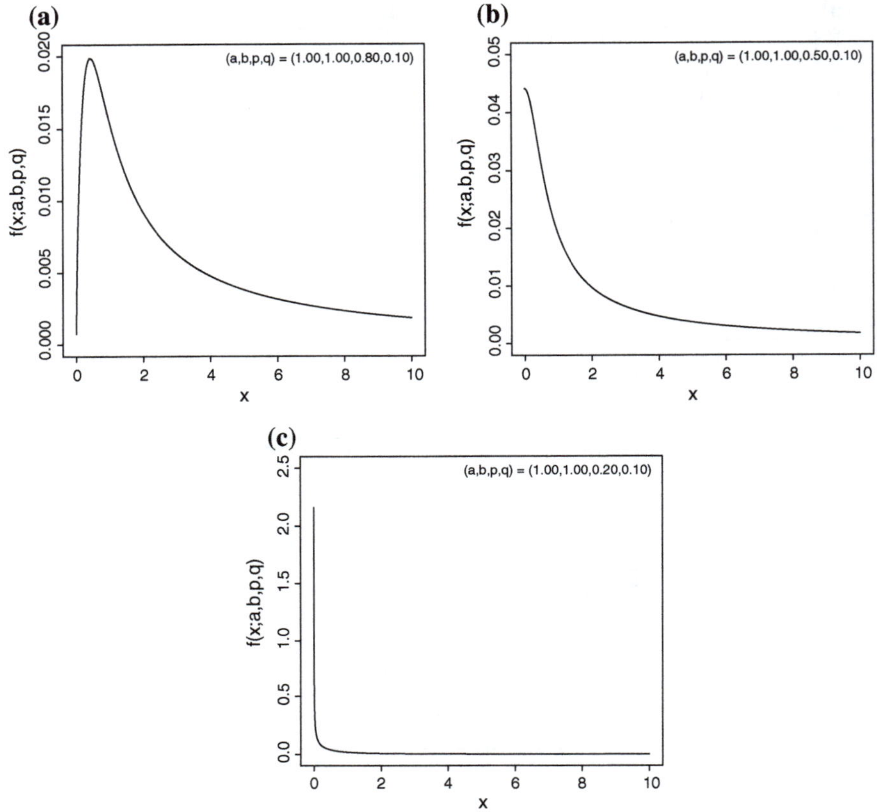

Fig. 5.4 IEκG2 density for various parameter values: **a** $p > \frac{1}{2a}$; **b** $p = \frac{1}{2a}$; **c** $p < \frac{1}{2a}$

Likewise, the tails of the IEκG2 follow the power-law forms

$$f(x; a, b, p, q) \underset{x \to 0}{\sim} c_7 x^{2ap-1} \quad \text{and} \quad f(x; a, b, p, q) \underset{x \to +\infty}{\sim} c_8 x^{-aq-1}. \tag{5.16}$$

The IEκG2 density approaches zero or a finite positive value or diverges to infinity when $x \to 0$ depending on whether, respectively, $p > \frac{1}{2a}$, $p = \frac{1}{2a}$ or $p < \frac{1}{2a}$. For $p > \frac{1}{2a}$, the density admits an interior mode; otherwise, it is always monotonically decreasing without further restrictions on the parameters—unlike the EκG2 density. Figure 5.4 depicts some IEκG2 densities.

5.2 Population Characteristics

Statistical tools for describing distributions and analyzing income inequality can be expressed analytically for the four-parameter variants of the κ-generalized distribution. Table B.6 reports formulas for the moments, the Lorenz curve, the Gini index, the coefficient of variation, the mean logarithmic deviation and the Theil index for each of the models discussed. These expressions are functions of the parameters in the respective models and are useful in analysis of population characteristics. Given parameter estimates, these results can be used to estimate corresponding population characteristics of interest as well as to provide indirect checks on the validity of the parameter estimates.

5.3 Empirical Results

The four-parameter models discussed in the present chapter were used by Okamoto (2013) to analyze household income/consumption data for approximately 20 countries selected from Waves IV to VI of the LIS database.[4] The data were grouped into 22 classes and 6 different variables (equivalized/non-equivalized household gross income, equivalized/non-equivalized disposable income and equivalized/non-equivalized consumption) were considered for each country data set. To provide a comparison with alternative four-parameter models of income distribution, the GB1 and GB2 (McDonald 1984) and the double Pareto-lognormal (dPlN) distribution introduced by Reed and Jorgensen (2004) were also fitted to the same data sets and welfare variables.[5] In all cases, parameter estimates were obtained using the maximum likelihood method for grouped data[6] and the goodness of fit assessed using both frequency-based (FB) evaluation criteria—such as the log-likelihood value—and money-amount-based (MAB) measures, which are expected to more closely reflect the accuracy of inequality estimates—some examples of MAB measures are the square root of the sum of squared errors between the observed and estimated Lorenz curve (LRSSE) and the absolute error between the observed and estimated Gini index (AEG).

[4] Here we limit ourselves to synthesize the main results obtained from a much detailed analysis by Okamoto (2013) and refer the interested reader to his paper for further details.

[5] We refer to the above-mentioned works for the basic properties of these models. Analytic expressions for the Gini index and the Lorenz curve of the GB1 and GB2 distributions are found, respectively, in Okamoto (2013) and Kleiber and Kotz (2003). Jenkins (2009) provides formulas for the mean logarithmic deviation and Theil indices within the GB2 family, while those of the GB1 distribution are given by Okamoto (2013). Finally, all expressions for inequality measures of the dPlN distribution can be found in Okamoto (2012a, b, 2013).

[6] Details on maximum likelihood estimation of income distributions from grouped data can be easily obtained, among other sources, via the volume by Kleiber and Kotz (2003), the collection of papers in Chotikapanich (2008), and papers by Bandourian et al. (2003) and McDonald and Xu (1995a, b).

Overall, the IEκG1, IEκG2 and GB1 resulted in a very poor fit in terms of both FB and MAB measures. The EκG1 is inferior to the EκG2, GB2 and dPlN in terms of the log-likelihood value, but superior to the EκG2 and GB2 according to the LRSSE and AEG in the overall evaluation. As for the comparison between the EκG1 and dPlN in terms of MAB criteria, the former appears to be outperformed by the latter in terms of both the LRSSE and AEG, except for the equivalized gross and disposable income variables.

No matter how the goodness-of-fit measures are combined, the pairwise comparisons between the EκG2 and GB2 show that the former clearly outperforms the latter in the vast majority of cases. The EκG2 is also dominant over the dPlN in terms of the log-likelihood value and the LRSSE for both the non-equivalized and equivalized income/consumption variables, whereas the AEG tends to slightly favor the dPlN—especially in the categories of non-equivalized variables.

On the whole, the EκG2 outperforms other four-parameter income distribution models. In particular, the EκG2 is dominant over its counterparts in almost all categories of equivalized variables. It is also superior to the EκG1, GB2 and dPlN in the categories of non-equivalized variables, although slightly inferior here to the dPlN when considering the accuracy of the estimated Gini index. Since the EκG2 can be viewed as a new kind of generalized beta distribution—close in form to the GB2, which is widely recognized as providing an excellent description of income distributions—this four-parameter variant of the κ-generalized distribution appears to drill a significant extension to the utility of parametric models for analysis of income distributions.

Chapter 6
Conclusions

I would like to use this occasion to give further impetus to the re-incorporation of income distribution into the main body of economic analysis.

Anthony Barnes Atkinson

Abstract This chapter draws the main conclusions from, and reflects on, the research presented throughout the book. Suggestions for future work are also proposed.

Keywords Income/wealth size distribution · Economic methodology · Complexity · Heterogeneity · Interaction · Probabilistic laws · κ-generalized distribution

Over recent decades, the distribution of economic resources among individuals and households has been deteriorating in many advanced and developing countries, leading to increased disparities between the rich and poor, vulnerability of the middle class and growing public demand for income and wealth redistribution (Atkinson 2015; Fitoussi and Stiglitz 2009; OECD 2008, 2011, 2015; Piketty 2014; Stiglitz 2012, 2015a). Furthermore, increasingly convincing evidence has been provided to show that deteriorating distribution has harmful effects on economic growth and tends to amplify the risk of crisis (e.g. Ostry and Berg 2011; Cingano 2014; Ostry et al. 2014; and Dabla-Norris et al. 2015). As a consequence, the issue of income and wealth distribution has featured prominently back in the economics research agenda, especially since the Great Recession (2008) and the consequent inability of the world economy to recover fully from the crisis. The availability of new and better data has attracted and facilitated studies into this issue in recent years, leading to a definitively renewed interest in the problems of income and wealth distribution. It seems therefore that the economics discipline is now willing to attempt at "bringing income distribution in from the cold", as demanded by Atkinson (1997) almost 20 years ago in his presidential address to the Royal Economic Society.

F. Clementi and M. Gallegati, *The Distribution of Income and Wealth*, New Economic Windows, DOI 10.1007/978-3-319-27410-2_6

6.1 The Main Results of this Book

Spurred by the resurgence of interest for distributional issues, this book has been devoted to illustrating new advances into the field of parametric modeling of income and wealth data. Attention was mainly drawn to the "κ-generalized" distribution, a new statistical model based on the κ-deformation of the standard exponential and logarithm functions that follows from Kaniadakis (2001, 2002, 2005). This model has a bulk very close to the Weibull distribution—which is recovered when the deformation parameter κ tends to zero—while for high values of income/wealth its upper tail approaches a Pareto distribution. Expressions for the shape, moments and various tools for inequality measurement were derived; given estimates of the parameters, these expressions can be used by income and wealth analysts to estimate corresponding population characteristics of interest as well as to provide indirect checks on the validity of the estimation procedure.

The performance of the proposed distribution, together with its extensions and generalizations, has been checked against household income/wealth data of different countries and years as provided by the Luxembourg Income Study (LIS) and the Luxembourg Wealth Study (LWS) databases. Parametric models belonging to the κ-generalized family appeared to provide a very good description over the entire income/wealth range, including the upper tail, and the inequality analysis expressed in terms of their parameters revealed very powerful. Furthermore, it was found that in the majority of cases the performance of these models is to be considered superior when compared to some alternative parametric forms for the size distribution of income and wealth.

There have been few attempts, to date, at developing an economic theory of the distribution of income and wealth among individuals. In particular, the shape of actual income and wealth distributions—i.e. why the upper tails of the distributions are Pareto and why at lower levels of income/wealth the distributions seem to be described by a different law—still lacks a satisfactory explanation. Some notable efforts made over the past 100 years either regarded existing distributions as the outcome of certain stochastic processes, (e.g. Gibrat 1931; Champernowne 1953; and Mandelbrot 1961) or as the result of the combined influence of decisive factors like home environment, talent, schooling and hierarchical position (Lydall 1968). However, these theoretical achievements have been only partial and piecemeal responses to the challenge of addressing the main stylized facts of the shape of personal income/wealth distribution (Sahota 1978; Stiglitz 1969).

6.2 Keys to Our Analysis

For so long, the subject of personal income/wealth distribution has been playing a peripheral role in economic analysis. Both classical economists and mainstream economic theory were certainly concerned with the determinants of payments to factors—labor, land and capital—but the relationship of the functional distribution with the distribution by size was typically not spelled out. In particular, the dominant paradigm in macroeconomics has been criticized for having long ignored (re-)distributional issues by reasoning within a modeling framework in which all distributional considerations are assumed away—the so-called "representative-agent" framework.

In representative-agent models, which have become ubiquitous in economic teaching and research since the 1960s, one assumes from the outset that all agents have identical preferences and resources—i.e. receive the same wage, are endowed with the same wealth and enjoy the same sources of income—so that growth proportionately benefits all social groups by definition. Probably, many macro-economists would justify the use of a representative-agent model on the ground that, while it is not precisely true that all the agents have identical endowments and tastes, it is nevertheless a close enough approximation. But in this case being "close" does not count. Indeed, as Kirman and Koch (1986; but see also Kirman 1992) show, in order to argue that the general equilibrium framework provide us with micro-foundations for macroeconomics, one must believe that is precisely true that all agents have the same endowments and preferences. Similarly, representative-agent models cannot address the consequences of changes to the distribution of income and wealth, and so they cannot address the impact of redistributive policies on inequality[1] or the impact of inequality on macroeconomic variables which is center stage in recent economic debate (Stiglitz et al. 2006).

It is no surprise, then, that mainstream economic theory has had little to say about the consequences of the most recent crisis for the distribution of economic welfare—or, conversely, how rising economic inequality somehow created macroeconomic instability. The historical moment we are going through has decisively uncovered flaws in the dominant view, and the revived concern around the distribution of income and wealth has helped in confirming that distributional analysis matters a great deal for thinking about the macro-economy.

Much of the discontent with contemporary economics has often materialized in public rejection of the main claims of modern mainstream economists, typically citing their failure to predict the economic crisis as a major reason for disregarding their opinions on contemporary matters. The issue, we believe, is broader, and involves questioning the current status of the discipline as a science.

Mainstream economics is a field where research is largely conducted with the language of mathematics—often incomprehensible to the uninitiated—and consequently many of those who feel strongly that there is something deeply wrong with this way of thinking economic analysis struggle to locate the precise point at which its arguments go astray. The basic paradigm of "standard economics", the general equilibrium theory, is a formal imitation of physics for both appearance and the epistemology: it comes up with clearly stated axioms, mathematical models for applications, systematic data collection and statistical analysis for empirical evidence.

[1] A representative agent has welfare significance only if lump-sum redistribution of endowments is possible (Atkinson and Stiglitz 2015).

However, although resembling physics and despite the alleged existence of "laws", economics is not a science.

Science can be defined as the body of knowledge obtained through the "scientific method", which consists of iterative applications of inductive and deductive methods to achieve consistency between the implications of a theory and observations. But most of the economic knowledge has not been obtained through the combined use of theoretical and experimental methods. A stable theory that is well tested to be consistent with observations, such as theories of physics, does not seem to exist in economics—and one may wonder if it ever will exist.[2]

Also, the nature of laws in economics is different from that of physical sciences: while the physical universe is governed by immutable laws in space and time, beyond the pale of human behavior, economics is a social science whose laws are influenced by human behavior—because economic agents, unlike atoms, learn from experience and have a will—so that things are seldom equal.[3] Simply put, we cannot change the speed of light no matter how we behave but we can change the distribution of income/wealth by our behavior. This explains why physicists just need to "describe" whereas economists always need to know the "how and why" of any case under their examination.

The short point is that economics cannot lay claim to the immutability, precision and exactitude of the laws of physics. Economic "laws" are rather assumptions or hypotheses which are very much a function of the context and often contradict the facts—although economists have long developed theories based on these alleged laws before there were sufficient systematic observational data to test their validity.

The reason why economics is not a science is due to the lack of interaction between the theoretical and empirical layers, which exist in parallel but rarely intersect to create real scientific knowledge.[4] As Kaldor (1972, p. 1239) noted:

---

[2]As already Marshall (1920, pp. 448–449) pointed out, economics has no near kinship with any physical science: "It is true that the forces with which economics deals have one advantage for deductive treatment in the fact that their method of combination is [. . .] that of mechanics rather than of chemistry. [. . .] But even in mechanics long chains of deductive reasoning are directly applicable only to the occurrences of the laboratory. By themselves they are seldom a sufficient guide for dealing with the heterogeneous materials and the complex and uncertain combination of the forces of the real world. For that purpose they need to be supplemented by specific experience, and applied in harmony with, and often in subordination to, a ceaseless study of new facts, a ceaseless search for new inductions".

[3] "[. . .] The matter with which the chemist deals is the same always: but economics, like biology, deals with a matter, of which the inner nature and constitution, as well as the outer form, are constantly changing" (Marshall 1920, p. 449).

[4]Econometrics seemed to possess the methodology that makes it possible to discover "scientific" laws of economics, similarly to the analysis of observational data in astronomy. Unfortunately, it has not been so: empirical research in econometric studies—mainly conducted by the econometric branch of mainstream economics—has widely scrutinized economic data to develop advanced techniques of mathematical statistics for testing hypotheses (e.g. normality of the errors, linear relationships, representative agent) without any significant contribution to economic theory. But this has not prevented mainstream economics to be regarded as "scientific", because of its willingness to consider empirical evidence. This created a scientific appearance ("illusion") for mainstream economics, elevated its status and enhanced its influence (Summers 1991).

> In economics, observations which contradict the basic hypotheses of prevailing theory are generally ignored: the "theorist" and the "empiricist" operate in two isolated compartments and the challenge of anomalous observations is ignored by the theorist—as something that could be taken into account at the stage of "second approximation" without affecting the basic hypotheses.

And needless to say, the dominant model in economics fails to prove empirically valid: the basic assumptions of perfectly rational and fully informed agents in a situation of general equilibrium have either no evidence or are contradicted by the best available evidence. However, the fact that almost all data contradict the assumptions of the model has been usually considered a minor problem compared to the advantage resulting from the fact that these assumptions can be handled mathematically to derive closed-form analytical solutions. But finding an analytical solution, although it may strengthen the model, is not decisive for the quality or less of the same; most of all, mathematical tractability can be in no way the final arbiter of science when it comes to model real-world systems.

For economic theories to be useful to the real world, economists should be prepared to modify or abandon their theories as necessary according to the facts of observation. To systematically apply the scientific method, the economics discipline needs a scientific revolution that knows how to tie the internal coherence to the external one within the borders of a social science that has many regularities but few laws.

6.3 Thinking About the Size Distribution of Income and Wealth

The position we took in the analysis presented throughout the book was that of "phenomenological relativism": the same phenomenon can be explained in different ways depending on the type or species of observation unit it refers to but, *mutatis mutandis*, the phenomena we are interested in have principles and minimal regularities that can be detected through societies and time. Then, we did not start from a theoretical system of internally consistent assumptions, the consequences of which must be observed in the manifestations of the phenomenon; rather, we have started from the observed phenomenon to reach a synthesis aimed at understanding the phenomenon in its fundamental aspects.

The major pillars of our analysis can be summarized as follows.

- At the elementary level, socioeconomic systems consist of *animate* beings which—unlike atoms—do what they want, if possible; they are also *thinking*, i.e. capable of making decisions, learning and discovering what is unknown to others, dominated by a *vital impulse* that drives them to improvement of their condition and satisfaction of their needs—also by simply imitating or adapting—and *reasoning*, that is unpredictable beyond any paradigm of rationality.
- Socioeconomic systems are complex because their constituents, regardless of the granularity level of description, are characterized by the categories of

heterogeneity and *interaction*, which feed off each other and overlap to originate *emergent phenomena*.

- Micro-, meso- and macro-quantities can all be *combined algebraically*, and accordingly are also *transferable*, but functional operators describing them can not. That is, a quantity can be represented by a stochastic operator, often non-linear, whose numerical realizations are summable but the different functionals are not, and this implies using an inferential method for micro-founding a macroscopic model.
- In reference to one or more state quantities, the *equilibrium condition of a complex system in evolution* is represented by a probability distribution approaching the stationary state in time, also consistent with a relative imbalance of its constituents which, because of their heterogeneity, are induced to interact so as to give rise to further heterogeneity.
- As a result of the emergence phenomenon, the evolutionary dynamics of complex systems can only be described by resorting to inferential principles of (a sort of) *stochastic determinism*, that is the probabilistic specification of functions that at the aggregate level represent the most likely trajectory the system will follow save for possible anomalies (fluctuations around the trend) induced by microscopic interactions.

One of the most appropriate methods to investigate complex systems is that of statistical mechanics, which directly connect the relevant microscopic information to useful macroscopic quantities by means of a probabilistic approach. The large number of interacting particles (agents) does not make it possible, indeed, to proceed with the deterministic method of classical mechanics (Khinchin 1948). Furthermore, complex socioeconomic systems constitute of agents which are not all the same as atoms of an ideal gas but heterogeneous with regard to both endowments and behavior. The continuous overlapping between heterogeneity and interaction generates income/wealth concentration and inequality.

When income/wealth flows/accumulates to people in a particular class, all members of this class are "identical" with respect to the class they are enrolled because of their level of income/wealth (*weak heterogeneity*) but not as compared to the way income or wealth enabling access to different classes has been gained (*strong heterogeneity*). Therefore, disparities cannot be explained only by relative differences in levels of income and wealth (weak heterogeneity); they also result from the way agents' action has been allowed to take steps toward those levels (strong heterogeneity). Put differently, in addition to talent and effort, which determine relative income/wealth positions, unequal opportunities—economic, political and social—that prevent agents to reach their full potential can also be held responsible for unequal outcomes.

Furthermore, as in physical collisions particles with very large energy tend to exchange only a small part of their energy when colliding with less energetic particles, in economic interactions the richest individuals tend to put at stake, in their interactions, only a small part of their income/wealth, because of their higher propensity to save. This enables them to protect their status and exclude the rest from joining their "club", leading to reinforcement of heterogeneity.

These considerations lead to an "exclusion" principle that calls for different probabilistic laws (*relativism*) depending on the part of the income/wealth support the analysis is focused on:[5] as in statistical mechanics many bosons obey the Bose-Einstein statistics at low energies and give rise to the family of exponential distributions, whereas few fermions obey the Fermi-Dirac statistics at high energies and follow a power law, at the bottom of the income (wealth) distribution many individuals earn (accumulate) little of it as compared to the gains of few individuals located at the top.

At the bottom of the income/wealth distribution no exclusion principle is at work: there may be as many poor (bosons) as you can have both at the time of the Pharaohs and today, with the only limit being the total number of units and total income or wealth. At the top, however, the exclusion principle is in force and becomes stronger the higher the social ladder is climbed, so that you can have a few CEOs (the strongest fermions) but only one Pharaoh (the extreme fermion). Whether obeying or not an exclusion principle determines the asymmetry of the income/wealth distribution and represents a form of relativism in such a way that the reference probabilistic principle is the family of exponential distributions for the bottom of its support, whilst for the top the reference probabilistic principle is the family of power-law distributions.

6.4 Where Do We Stand and Where Are We Headed?

Among the many parametric models for the size distributions of income and wealth proposed in the literature, the κ-generalized model is the only one which formally and explicitly embeds both the exponential and the power-law shapes in a single functional form: therefore, when operating on the portion of the income/wealth distribution ruled by probabilistic principles of the exponential family (no exclusion principle) the κ-generalized model replicates remarkably well the underlying data, and the same good performance holds when it operates on that portion of the income or wealth support ruled by probabilistic principles of the power-law family (exclusion principle), thus capturing both the mentioned relativistic aspects.

Hence, the κ-generalized model reveals very powerful because it is not pure formalism but almost pure phenomenology: that is, a probabilistic model where event and probability spaces interact so as to grasp the probabilistic characters of the phenomenon for the description of which it is used. We believe this is a fundamental logical progression in the way of reasoning about economic quantities such as income or wealth and their distribution among the population.

Of course, we are a still a long way off from the end of the story. What is actually missing in order to proceed in the field is a theoretical model able to demonstrate the emergence of κ-generalized income/wealth distributions as the result of decentralized interactions of a large number of heterogeneous agents. Mainstream economic theory, centered around the Walrasian methodology and the representative agent assumption, is definitely badly equipped for this task. On the contrary, agent-based

[5] This principle is examined in detail in Sect. A.6 of Appendix A.

computational economics—i.e. the use of computer simulations to grow and study evolving artificial economies composed of many heterogeneous interacting agents—represents a promising tool for advancements along this field of research that we expect to address the new theoretical challenges on the size distribution of income and wealth raised by Stiglitz (2015b, c, d, e). Since our efforts will be directed in this direction from now on, provided he has survived so far we hope the reader will also have the patience to wait and retrieve our future results.

Appendix A
Mathematics of *Strange* Quantities: Why Are κ-Generalized Models a Good Fit to Income and Wealth Distributions? An Explanation

The aim of science is to seek the simplest explanations of complex facts. We are apt to fall into the error of thinking that the facts are simple because simplicity is the goal of our quest.
Alfred North Whitehead

A.1 A Narrative Motivation

While developing a relativistic interpretation of statistical mechanics, Kaniadakis (2001) introduced the κ-distribution with further developments and details proposed by Kaniadakis (2002, 2005).[1] The κ-distribution, named *κ-generalized distribution* after Clementi et al. (2007), has been successfully applied in physics but also in economics, mainly for modeling the size distributions of income and wealth (see Clementi et al. 2007, 2008, 2009, 2012a, b). But while physics applications of the κ-generalized model receive a clear and exhaustive interpretation of all its components, in economics things are not so clear; in a sense: it works well but it is still not well understood why. It is therefore the aim of this Appendix to explain the reasons *why* the κ-generalized model works so well while discussing the case of income and wealth distributions. Also, it will provide a conceptual and methodological bridge to the technical chapters with applications, where details about *how* the κ-generalized model works are given.

The whole reasoning is grounded on a weak analogy with the special relativity and develops the notion of *strange quantity*, that is analyses the peculiarities driving (economic) quantities away from being *regular*. The main thesis is twofold: on the

[1] This appendix was written by Simone Landini—IRES Piemonte (Socio-Economic Research Institute of Piedmont, Turin, Italy). He is grateful to Fabio Clementi and Mauro Gallegati for their invitation to develop this note and Giorgio Kaniadakis for fruitful conversations. Views and opinions are those of the author and do not involve the responsibility of IRES Piemonte.

F. Clementi and M. Gallegati, *The Distribution of Income and Wealth*, New Economic Windows, DOI 10.1007/978-3-319-27410-2

one hand, it argues that the κ-generalized is the appropriate probabilistic model for strange quantities; on the other hand, it also argues that almost all the quantities relevant to economics are characterized by a strange phenomenology which finds a clear interpretation under the light of complex systems paradigm. Moreover, it is a further conjecture that, beyond the economic horizon, many complex systems quantities behave strangely.

The notion of strange quantity is composed of several aspects. A strange quantity is *detectable at any level of granularity*, from the very macro level of the system as an ensemble, say $X(t)$, to the very micro one of its constituents, say $X(\omega, t)$, whose behaviors determine the emergence of systemic properties not directly addressable to any part ω of the system Ω but to their superimposition. That is, the phenomenology of a strange quantity can be properly understood with the categories of complexity such as heterogeneity, interaction and emergence. In addition, a strange quantity is *algebraically aggregative*—or additive, i.e. $X(t) = \sum_{\omega} X(\omega, t)$emdash even though the systemic value cannot be disaggregated other than inferentially. This means making inference about a probabilistic model $X \sim \mathcal{K}(\boldsymbol{\theta})$ which is maximally consistent with the structural properties of the system to determine with minimal uncertainty the configuration generating the aggregate outcome: this aspect already anticipates that the Jaynes' (1957a, b) maximum entropy principle and the statistical mechanics perspective provide the appropriate methodological framework.

Along these lines, several economic quantities can be told to be strange. The main motivation for this conjecture can be traced by observing the stylized facts which can be exhaustively explained in terms of a probabilistic description. As a consequence, a strange quantity is better understood as a *random variable*, a *stochastic process* or a *random field*. The essential stylized facts emerging from the structure of the system concern, among others, the *concentration* and *inequality* of the distribution, both due to the behavioral superimposition of many *heterogeneous* constituents repeatedly *interacting* by exchanging slots of a transferable quantity X without influencing its total amount, which may even grow in the long run while being almost constant in the short run. Such stylized facts can be explained by means of a unified probabilistic model $\mathcal{K}(\boldsymbol{\theta})$ which includes two families of distributions: the family $\mathcal{E}(\boldsymbol{\theta})$ of *exponential distributions*, to grasp concentration effects around the expected value, and the family $\mathcal{P}(\boldsymbol{\theta})$ of *power-law distributions*, to grasp inequality due to long tails. The coexistence of these two different probabilistic principles characterizes a strange quantity.

The probabilistic model for the distributions of income and wealth described through the chapters of this book operates such a synthesis both by means of *numerical transformations*, to make the effects of heterogeneity and interaction more intelligible, and by means of *functional deformations*, to model concentration and inequality. The κ-generalized model is indeed able to jointly describe the distribution both where some principles rule the concentration of the system, say $\forall X(\omega, t) \leq \xi$ then $X \sim \mathcal{E}(\boldsymbol{\theta})$, and where other principles rule the inequality of the system, say $\forall X(\omega, t) > \xi$ then $X \sim \mathcal{P}(\boldsymbol{\theta})$. Put differently, the κ-generalized distribution is the appropriate probabilistic model for strange quantities described by random variables whose probability space is not isotropic with respect to partitions of the state space.

In what follows, this narrative will be formally developed but before mathematical subtleties a digression about relativity is worth developing to explain the underlying philosophy: although sacrificing lots of formal aspects, a weak analogy with the relativity theory is discussed to introduce how Kaniadakis (2001, 2002, 2005) obtained the κ-generalized model while developing a relativistic interpretation of statistical mechanics.

The fundamental principle of relativity (FPR) asserts that physics laws are the same in every homogeneous system of reference. The postulate of the Galilean relativity (GR) is that the laws of mechanics do not change on different inertial systems of reference reciprocally moving with uniform and linear motion. By means of Galileo transformations an observer in a system of reference can deduce what another observer has observed just knowing the motion speed of two inertial systems: Galilean transformations are translation symmetries in the Euclidean space, which is homogeneous and isotropic, to be applied to macroscopic phenomena at sub-luminal velocities.

The Einstein relativity (ER), namely the special relativity (SR), starts from the GR postulate, extends it from the mechanics to electromagnetism and optics, and introduces a new postulate: light signals propagate at a constant velocity c not depending on the source and the receiver. The difference between GR and ER takes care of the speed of light as a physical limit: while for the GR simultaneous events with respect to an observer are simultaneous also with respect to another one in a different system of reference moving with linear and uniform motion at sub-luminal velocity, for the ER simultaneous events for the first observer can be non-simultaneous for the second if bodies move at luminal velocities. Furthermore, not only time dilatation but also space dilatation is possible.

In this scenario, to translate measurements from one system to another it needs using the Lorentz transformation, which finds application in the four dimensions Minkowski space-time, where distances between events follow a *deformation* of the Pythagoras' theorem due to time and the speed of light. Therefore, in the Galilean world space and time are distinct and absolute categories while in the Einstein one are entangled and relative, which is quite *strange*. Moreover, the ER finds the GR as a particular case, i.e. when velocities are sub-luminal—on the contrary, ER phenomena cannot be observed in the frame of the GR, because the velocities are about that of light. Roughly, but intuitively enough, this is what needed to understand that space-time coordinates of the ER depend on c as a deformation parameter introduced by the Lorentz transformation. As it will be shown, something similar happens in the specification of the κ-generalized model, which depends on a deformation parameter κ. As depending on c the ER finds the GR, depending on κ the κ-generalized model can grasp the exponential or power-law behavior of the strange quantity X on its support. In a sense, the deformation parameter κ allows the κ-generalized model to describe the income or wealth distribution according to an exponential family distribution when X is below a given threshold ξ and according to a power-law family distribution when X is above ξ: interestingly enough, the κ-generalized model is a single distribution unifying both shapes at once.

A.2 *Minimalia*

The very elementary notion to begin with is that of *observation unit*, ω, i.e. that entity upon which one observes the degree of absence or presence of a property related to a phenomenon of interest.

An *observable*, $\mathcal{X}$, is a property characterizing an observation unit like, for instance, the richness of a person. An observable such as richness is methodologically said to be a *quantity*, X, to mean income or wealth or else, once an *operational definition* is provided. This means that an entity becomes relevant to knowledge with respect to a given observable if such a property can be somehow measured:[2] at least in quantitative sciences, *observing means measuring*, which is a complicated process. First of all, it needs defining a unit of measurement, detectable by means of a detector specific for the quantity associated to an observable. Then, it needs counting how many times the chosen unit of measurement belongs to the observation unit, which requires a reference unit of measurement.

A *reference unit of measurement* (RUM), Δ, is an amount, a mass or a mole conventionally chosen as the term of reference for measurement of an observable. Put differently, the RUM of an observable is the conventional term of reference allowing for observation of the degree at which a given observation unit is characterized by a specific observable property.

For instance, saying that Shirley is rich it does not tell how much income she earns or wealth she owns: this is not measurement but classification, as it depends on where one places the threshold between rich and poor. That is, she may be rich with respect to another one who earns less money but she may be anyway poor with respect to the one earning a higher amount of money, although they are all above the threshold between rich and poor. Therefore, measuring Shirley's richness requires more detailed information about her income—of course if one refers to income as the appropriate quantity to identify and analyse richness as property.

Furthermore, assume $\Delta = 1\$$. Saying that Shirley's income is 57,385$ is a detailed measurement of her income, which informs about her state of richness. This is because there is a RUM conventionally set at $\Delta = 1\$$ according to which she owns 57,385 times above 0: if Shirley were earning an income of 0$ then she would be poor because, as an observable, the property of richness is absent for Shirley (the observation unit).

Hence, the *measurement* $X(\omega)$ of an observable $\mathcal{X}$ is the outcome of a process ascertaining an amount that quantifies the number of times the chosen RUM of the quantity X is possessed by an observation unit ω in order to evaluate the degree at which the latter is characterized by the property $\mathcal{X}$.

This reasoning leads to a few remarks. First of all, measuring requires a RUM and an absolute point of reference: in the previous example the absolute point of reference was the level of 0$ income, which is the lowest bound for income as a quantity. Secondly, the more the measurements $X(\omega)$ of the quantity X associated to the observable $\mathcal{X}$ are detected near the lower bound, the less those observation

[2]Usually, "observable" and "quantity" are terms used as synonymous.

units ω posses the property to which the quantity is associated to: roughly, the lowest the income the less the richness. Finally, beyond richness and income, it becomes clear that it is not always easy, as far as possible, making lower and upper bounds explicit, mainly because the economy is stimulated to growth. Nevertheless, it is a matter of fact that *no quantity which is relevant to economics may be unbounded*, neither from below nor from above. Hence, it would be a good practice recognizing that any measure $X(\omega)$ is bounded between $x_{\min}$ and $x_{\max}$. Of course, it is not always easily possible to define a set $\mathbb{X} = [x_{\min}; x_{\max}]$ for $X(\omega) \in \mathbb{X}$, but it nevertheless exists—although it dynamically may change—and hence it should be fixed as arbitrarily or conventionally as possible on the ground of available information.

Therefore, measuring an observable $\mathcal{X}$ upon an observation unit ω means finding out to what extent ω is characterized by that phenomenon semantically connected to the observable through the operational definition of the quantity X. In the context of income and wealth, the measurement $X(\omega)$ of $\mathcal{X}$ upon ω is the information detected by means of an apparatus (e.g. a census, a survey or an interview) quantifying how many times the RUM $\Delta = 1\$$ is possessed by ω to know its state of richness.

A.3 Traces of *Strangeness*

Consider a set $\mathbb{X} = \{x \in \mathbb{R} : -\infty < x_{\min} \leq x \leq x_{\max} < +\infty\} \subset \mathbb{R}$ and a sample $\Omega_n = \{\omega \in \Omega\}$ of $n = |\Omega_n|$ observation units from the population Ω. Define a function $X : \Omega_n \rightarrow \mathbb{X}$ such that $(\omega) \mapsto X(\omega)$. If Ω_n were a sample of n earners, then $X(\omega)$ would be income.

From here on it shall be assumed that X is a random variable representing income or wealth, or every other quantity endowed with a RUM. Moreover, as income, X is a monetary quantity in the currency $\mathbb{D}$ and it is *transferable*: that is, a portion of income $X(\omega)$ of ω can be transferred into the income $X(\omega')$ of ω' without changing of total income in the economy.

Not depending on the currency, X is a discrete quantity *by nature*. This means income transfers among individuals come in lumps of arbitrary size because X can be measured with absolute precision. Discreteness of monetary and transferable quantities is implied by the existence of a *conventional* RUM (CRUM), such as $\mathbb{D} = \$$, representing the (unit of a) currency as a multiple of an *elementary* RUM (ERUM), such as $\delta = 0.01\mathbb{D}$, that is the ¢ of a \$: the ERUM of a CRUM is a RUM which can not be divided into smaller meaningful pieces. Hence, that of the ERUM is the highest level of precision for monetary measurements, i.e. the absolute precision can be reached if the chosen RUM is the ERUM $\Delta = \delta$ of X, and this is the reason why a quantity X endowed with an ERUM is intrinsically discrete by nature.

The *dimensional equation* of $\mathcal{X}$ explains the chosen RUM. If the RUM is the CRUM $\Delta = \mathbb{D}$, then the dimensional equation is $[X] := \mathbb{D}$; if the RUM is the ERUM $\Delta = \delta$, then $[X] := \delta$ (clearly both are equivalent because $\mathbb{D} = 100\delta$). That is, if one measures income in dollars, then $[X] := \$$;if one measures income in cents, then $[X] := ¢$, where $\$ = 100¢$. Therefore, in general, $[X] := \mathbb{D} = 100\delta$ and $[X] := \delta = 0.01\mathbb{D}$ are precisely the same.

When income data are published in statistical reports they usually refer to income classes and are evaluated as sums, averages, medians, percentiles or quantiles. In this case they are available in multiples of $\mathbb{D} = \$$, say $10^3\mathbb{D} = 10^5\delta$. When income data are fine-grained at the observation unit level of individuals they may be reported with decimals of $\mathbb{D} = \$$, but usually they are approximated at the unity. For instance, if Shirley's income is $57,385.62\mathbb{D} = 5,738,562\delta$ the income values adjacent to Shirley's one are $57,385.61\mathbb{D} = 5,738,561\delta$ from below for, say, Wendy and $57,385.63\mathbb{D} = 5,738,563\delta$ from above for, say, James. Notice that, in terms of δ, both are integers, and this is not a negligible aspect. Indeed one may ask: is there any possibility for a fourth Manson's income to lye between Wendy and Shirley's or between Shirley and James' incomes? No. This is the consequence of discreteness, which extends to all quantities endowed with an ERUM $\delta = ¢$ as the minimal fraction of the CRUM $\mathbb{D} = \$$, that is all the monetary and transferable quantities receive a discrete or quantized description and, in principle, they can be measured with absolute precision. This example suggests the idea that one might find some troubles in describing a huge and heterogeneous sample of incomes using a continuous distribution defined on a *real* support $\mathbb{X} \subset \mathbb{R}$ when data are discrete by nature.

Every measurement $X(\omega)$ at the ERUM precision level of $\delta = 0.01\mathbb{D}$ is decomposable into two parts: the integer part r and the decimal part m. Then, it necessarily follows that $m \in [0; 99] \subset \mathbb{N}$, because $100\delta = 1\mathbb{D}$. For what concerns the integer part r, there is no precise limit on $\mathbb{Z}$ other the one implied by the fact that $X(\omega) \in \mathbb{X} = [x_{\min}; x_{\max}] \subset \mathbb{R}$. Hence, it must be noticed that $\mathbb{X} \subset \mathbb{R}$ is a very generic and weak statement because only some real numbers are eligible to represent realizations of $X(\omega)$. As a consequence, $\mathbb{X}$ is not only limited and closed but also finite and numerable due to its intrinsic ERUM. Stated otherwise, every quantity ruled by an ERUM is intrinsically *digital* and defined on a limited and closed support $\mathbb{X}$ embedded into $\mathbb{R}$, even though for the sake of technical simplification it may be modeled as if it were *analog* on $\mathbb{R}$.

Therefore, even though $\mathbb{X}$ is embedded into $\mathbb{R}$, the very real *infinitesimal* cannot be rigorously defined, and this does not allow for properly involving infinitesimal calculus. Nevertheless, when operating on the orders of magnitude of $\mathbb{X}$, which is discrete by nature, infinitesimal calculus formalism is involved as if the realization field $\mathbb{X}$ were a proper subset of $\mathbb{R}$. Roughly, it needs measuring X at higher orders of $\mathbb{D}$, such as billions, to approach the small δ as if it were an infinitesimal quantity. In other words, while operating upon the orders of magnitude of X one can resort to orders of magnitude higher than the CRUM $\mathbb{D}$ in such a way that values $X(\omega) \in \mathbb{X}$ were sufficiently dense to forgive that $\mathbb{X}$ is truly discrete. To go through with this issue numerical examples may help.

Let X be income, with $[X] := \mathbb{D}$, such that

$$\begin{aligned} X(\omega) = x\mathbb{D} = \left(r + \frac{m}{10^2}\right)\mathbb{D} = r\mathbb{D} + m\delta = 10^2 r\delta + m\delta \\ = \left(10^2 r + m\right)\delta \in \mathbb{Z}_+. \end{aligned} \tag{A.1}$$

For instance, if $x = 317,245.71$, then $r = 317,245$ and $m = 71$, $X(\omega) = 31,724,571\delta$ is a positive integer.

Let now $X(\omega)$ be in thousands of $\mathbb{D}$, i.e. $[X] := 10^3\mathbb{D}$, such that

$$X(\omega) = \left(\frac{x}{10^3}\right) 10^3\mathbb{D} = \left[\frac{1}{10^3}\left(r + \frac{m}{10^2}\right)\right] 10^3\mathbb{D} = (10^2 r + m)\delta \in \mathbb{Z}_+. \quad \text{(A.2)}$$

As expected, nothing has changed because the measurement is always the same in terms of the ERUM, while for the CRUM the difference only concerns the numerical representation. Indeed, the higher the order of the CRUM, the more the decimal places: if $[X] := 10^3\mathbb{D}$, then $X(\omega) = 317.24571 \times 10^3\mathbb{D}$ instead of $X(\omega) = 31,7245.71 \times \mathbb{D}$. Along with this observation one can then represent $X(\omega)$ with $[X] := 10^6\mathbb{D}$ to get $X(\omega) = 0.31724571 \times 10^6\mathbb{D}$, but in terms of the ERUM $\delta = 0.01\mathbb{D}$ the value would be always the same. Therefore

$$\begin{aligned} X(\omega) &= 317,245.71 \times \mathbb{D} = 317,245.71 \times 10^2\delta, \\ \text{(thousands)} &= 317.24571 \times 10^3\mathbb{D} = 317.24571 \times 10^5\delta, \\ \text{(millions)} &= 0.31724571 \times 10^6\mathbb{D} = 0.31724571 \times 10^8\delta, \\ &= 31,724,571 \times \delta. \end{aligned} \quad \text{(A.3)}$$

Next, consider $X(\omega)$ as being an average income along with a lower-level income, say $X(\omega_0) = 15,325.12\mathbb{D} = 1,532,512\delta$. The proportion between $X(\omega_0)$ and $X(\omega)$ in terms of δ is about 1 to 21. The equivalences (A.3) now read as

$$\begin{aligned} X(\omega_0) &= 15,325.12 \times \mathbb{D} = 15,325.12 \times 10^2\delta, \\ \text{(thousands)} &= 15.32512 \times 10^3\mathbb{D} = 15.32512 \times 10^5\delta, \\ \text{(millions)} &= 0.01532512 \times 10^6\mathbb{D} = 0.01532512 \times 10^8\delta, \\ &= 1,532,512 \times \delta. \end{aligned} \quad \text{(A.4)}$$

Finally, consider $X(\omega_1) = 43,957,620,931.76\mathbb{D} = 4,395,762,093,176\delta$ as a *top income*. The proportion between $X(\omega_0)$ and $X(\omega_1)$ is about 1 to 2,868,338, that is: it takes 2 millions 868 thousand and 338 people like ω_0 to match with a rich such as ω_1. Suggested equivalences then read as

$$\begin{aligned} X(\omega_1) &= 43,957,620,931.76 \times \mathbb{D} = 43,957,620,931.76 \times 10^2\delta, \\ \text{(thousands)} &= 43,957,620.93176 \times 10^3\mathbb{D} = 43,957,620.93176 \times 10^5\delta, \\ \text{(millions)} &= 43,957.62093176 \times 10^6\mathbb{D} = 43,957.62093176 \times 10^8\delta, \\ &= 4,395,762,093,176 \times \delta. \end{aligned} \quad \text{(A.5)}$$

Therefore, the three sample incomes are

$$
\begin{aligned}
X(\omega_0) &= 0.01532512 \times 10^8 \delta, \\
X(\omega) &= 0.31724571 \times 10^8 \delta, \\
X(\omega_1) &= 43,957.62093176 \times 10^8 \delta.
\end{aligned}
\tag{A.6}
$$

Hence, if the income of ω sounds high w.r.t. that of ω_0 it is anyway almost nothing w.r.t. to that of ω_1. And basically this is what happens with income samples: there is a few of low incomes toward the minimum, a considerable bunch of medium-high ones around the average, and there are very few top incomes toward the maximum which, nevertheless, absorb a considerable share of total income in the economy. Moreover, evidence shows that there is a portion of the realization field, manly around the average, where differences are gradual between adjacent values, while there are regions, mainly along the tails, for which such differences are huge. Said differently, there are regions of the realization field $\mathbb{X}$ which appear to be more *concentrated* than others which are more pulverized. Beyond this, it can be observed that the decimal representation becomes in a sense more *dense* as the order of magnitude in terms of the ERUM increases, but this essentially depends on the *inequality* of income distribution and on the width of the realization field $\mathbb{X}$, i.e. the distance between the top incomes and the absolutely poor ones.

On the field $\mathbb{R}$ numbers can be represented with an indefinite sequence of decimal places, hence a compact set in $\mathbb{R}$ is a not numerable infinity of numbers: this allows for rigorously introducing the infinitesimal quantity $\mathrm{d}x$, which is far beyond the ERUM δ. For practical purposes one can anyway consider the field $\mathbb{X}$ to be partitioned into a large number of small pieces, say 10^{11}, which means values of income X are evaluated as $[X] = 10^9\mathbb{D} = 10^{11}\delta$. Accordingly, this tantamount to assuming that the highest possible income is of the order of 10 billions of $\mathbb{D}$.

Since $\mathbb{X}$ is limited and closed between $x_{\min}$ and $x_{\max}$, one may assume $x_{\min} = 0$ and $x_{\max} = 10^{11}$ to rescale observations as

$$
\begin{aligned}
X(\omega_0) &= 0.00001532512 \times 10^{11} \delta, \\
X(\omega) &= 0.00031724571 \times 10^{11} \delta, \\
X(\omega_1) &= 43.95762093176 \times 10^{11} \delta.
\end{aligned}
\tag{A.7}
$$

Also, a widely used rescaling is the so called *max-min transform*

$$
\tilde{X}(\omega) = \frac{X(\omega) - x_{\min}}{x_{\max} - x_{\min}} \in [0; 1], \tag{A.8}
$$

which requires the extrema of $\mathbb{X}$ to be known or, at least, arbitrarily fixed. This is not a problem: it can always be done so as to map $\mathbb{X}$ into [0; 1], which seems to approach a better representation of X in terms of real numbers—although on the scale of the unitary interval. Moreover, if $x_{\min} \approx 0$ then $\tilde{X}(\omega) \approx X(\omega)/x_{\max}$. Accordingly, distances between very different incomes only change their numeric value on a different scale but they may still remain huge in the new scale. Therefore, in the end, beyond the scale nothing has changed.

There are other order-preserving numerical transformations mapping $\mathbb{X}$ into an equivalent set which appear to be much more dense to approach the structure of $\mathbb{R}$—although without matching it perfectly. The main candidate is the (natural) logarithm, which is a monotonic transformation that preserves the ordering. The logarithm also introduces a non-linear transformation defined only for strictly positive values, i.e. it requires $x_{\min} > 0$. Furthermore, by definition, it operates more strongly on huge values. For instance, consider $\ln[X(\omega_0)] = 14.24$, $\ln[X(\omega)] = 19.58$ and $\ln[X(\omega_1)] = 29.11$. Assuming that $x_{\min} = 1$ and $x_{\max} = 10^{13}$ in terms of the ERUM δ, hence $\ln(x_{\min}) = 0$ and $\ln(x_{\max}) = 29.93$, it follows from (A.8) that $\widetilde{\ln[X(\omega_0)]} \approx 0.48$, $\widetilde{\ln[X(\omega)]} \approx 0.65$ and $\widetilde{\ln[X(\omega_1)]} \approx 0.97$. As it is clear, the order has not changed but the max-min transform applied to the log-transformed values non-linearly maps $\mathbb{X}$ into (0; 1) in such a way that the income *masses* become closer to each other on the same space. Nevertheless, although their single values are more perceptible to human brain for comparison, their distances are harder to figure out in terms of the ERUM δ. That is, what is *big* in the *real* world is still *big* on the *log* world, although at a different scale.

In sum, the moral of the story is that one can "torture" strange observables by rescaling transformations as he pleases but they nevertheless would not change their strange shape. Hence, as discussed below, a different point of view is needed, that is: beyond numerical transformations of the realization field, to accommodate for the strange shape of data and their distribution it needs introducing deformations and a suitable mathematics.

A.4 The Exponential Family Hypotheses

As observed, the ERUM induces the most fine grained homogeneous partition of the support $\mathbb{X}$, which is a collection of isolated points each equidistant from adjacent ones by δ. Each of such points is a number valued at the ERUM level as $x_h = c_h\delta$, where $c_h \in \mathbb{N}$ is a *pure* number. Then, $\{c_h\}$ is an ordered sequence, and so is the sequence $\{x_h\}$ of feasible incomes. It follows that the finest grained partition is

$$\pi(\mathbb{X}; \delta) = \{x_h : h = 0, 1, \ldots, H\}, \tag{A.9}$$

where

$$\begin{cases} x_h = (c_0 + h)\,\delta = x_0 + h\delta, \\ x_0 = x_{\min}, \quad x_H = x_{\max}, \quad V_{\mathbb{X}} = x_H - x_0 = H\delta \gg 0, \\ H = V_{\mathbb{X}}/\delta. \end{cases} \tag{A.10}$$

Being this the most fine-grained partition of $\mathbb{X}$, the $n \ll H$ sample of incomes $X(\omega)$ may assume a restricted number of realizations $X(\omega) = x_h$; hence, there may be portions of $\pi\left(\mathbb{X};\delta\right)$ with a high density of observation units while others may be almost pulverized. In this case the support is not homogeneously concentrated or populated: as regards income, for instance, evidence shows that the higher is x_h beyond a certain level, the less the earners on such levels.

To have a sounder description, one may consider the possibility of grouping data into classes. In the absence of other information, the simplest approach to reduce the high heterogeneity of the finest grained partition $\pi\left(\mathbb{X};\delta\right)$ is to use equispaced intervals. After Sturges (1926), in the literature there have been several suggestions for binning a set into classes, all of them being based on specific hypotheses related to specific purposes. In the absence of other information but $\mathbb{X}=[x_{\min};x_{\max}]$, one may then consider dealing with the simplest hypothesis of working out a homogeneous partition, but how many intervals?

Suppose $x_{\min}=10^4\delta$ and $x_{\max}=10^8\delta$. Then $V_{\mathbb{X}}=10^4\left(10^4-1\right)\delta$. The *order of magnitude* operator is

$$I=\lfloor V_{\mathbb{X}};\delta\rfloor_{10}=\left\lfloor \log_{10}\left(\frac{V_{\mathbb{X}}}{\delta}\right)\right\rfloor=\left\lfloor \log_{10}(H)\right\rfloor, \tag{A.11}$$

which is the nearest integer to the base-10 logarithm of a positive number. In this example, therefore, $I=8$. Define then a *partition-constant*

$$\nu=\nu\left(V_{\mathbb{X}};I\right):=\left\lfloor\frac{V_{\mathbb{X}}}{I}\right\rfloor, \tag{A.12}$$

from which follows that $\nu=1,250\times 10^4\delta$. Notice that ν depends only on the knowledge of $\mathbb{X}$, which is the only available information. According to ν one may then specify the bins

$$x_i=x_0+i\nu, \tag{A.13}$$

with $i=0,\ldots,I$. Of course, $x_0=x_{\min}$ while $x_I=x_{\max}$. Accordingly, the intervals

$$A_i=\left[x_i;x_{i+1}\right),\quad i=0,\ldots,I-1, \tag{A.14}$$

induce a *homogeneous partition* of $\mathbb{X}$ into not-overlapping classes A_i

$$\Pi\left(\mathbb{X};\nu\right)=\{A_i:i=0,\ldots,I-1\} \tag{A.15}$$

such that

$$\bigcup_{i=0}^{I-1}A_i=\mathbb{X}\wedge\bigcap_{i=0}^{I-1}A_i=\emptyset. \tag{A.16}$$

Therefore, a measurement for X upon ω may realize in $\Pi\left(\mathbb{X};\nu\right)$ as a *measurement-event* $X\left(\omega\right)\in A_i$. Each A_i may now host several realizations $X\left(\omega\right)$, and this should reduce the initial pulverization because $\Pi\left(\mathbb{X};\nu\right)$ is now a gross-grained partition.

Define now a function mapping $\Pi(\mathbb{X};\nu)$ into a discrete set $\Lambda(\mathbb{X};\nu)$ of values $l_i = l(A_i)$. The simplest function one may deal with is the central or median value function

$$\Lambda(\mathbb{X};\nu) = \left\{ l_i = \frac{x_i + x_{i+1}}{2} = x_i + \frac{\nu}{2} = \left(x_0 + \frac{\nu}{2}\right) + i\nu \right\}, \tag{A.17}$$

$i = 0, \ldots, I-1$, for which l_i is a *sample-value* within A_i to synthesize all those $X(\omega) \in A_i$. This means A_i is an *equivalence class*

$$X(\omega) \in A_i \Leftrightarrow X(\omega) \doteq l_i, \quad \forall \omega \in \Omega_n. \tag{A.18}$$

Accordingly, $\Lambda(\mathbb{X};\nu)$ is called the *measurement-event field* induced by the gross-grained partition $\Pi(\mathbb{X};\nu)$ associated to the realization field $\mathbb{X}$. Notice that observation units in the class A_i are equivalent w.r.t. to measurements of X but they may be heterogeneous w.r.t. to other properties different from X. Moreover, the wider is $\mathbb{X}$ the huger is I, then the less $\Pi(\mathbb{X};\nu)$ is gross-grained and the huger is the number of values in $\Lambda(\mathbb{X};\nu)$: according to (A.12), everything still depends on $\mathbb{X}$ as the only available information.

Define then

$$q(l_i) = \frac{\#\left\{X(\omega) \doteq l_i : \omega \in \Omega_n\right\}}{n} \geq 0 \tag{A.19}$$

as the *density-indicator* for the measurement-event $l_i = l(A_i)$, a function which evaluates the share of observations belonging to A_i. Of course, $q(l_i)$ is an empirical estimate of a probability as the image of the event $X(\omega) \in A_i$ on the *density-field* $\mathbb{I} = [0; 1]$; its value can be estimated only if all the incomes $X(\omega)$ are known: if this is not the case, one can only make some inference about it.

As known, when no other information is available the best one can do to evaluate $q(l_i)$ is to assume equi-probability of measurement-events:[3] i.e. $q(l_i) = q$ for each $l_i \in \Lambda(\mathbb{X};\nu)$ associated to $A_i \in \Pi(\mathbb{X};\nu)$. Thus, the problem now is how to estimate q. Due to homogeneity of $\Pi(\mathbb{X};\nu)$, the width of A_i is equivalent to that of any other A_j, i.e. $x_{i+1} - x_i = x_{j+1} - x_j = \nu$, as well as the distance between mid-points l_i and l_{i+1} is constant at ν on $\Lambda(\mathbb{X};\nu)$. The value of q may then be estimated as the ratio between the width of A_i and the width of $\mathbb{X}$ associated to every event (A.18) by means of (A.12), that is

$$q = \frac{x_{i+1} - x_i}{x_I - x_0} = \frac{[x_0 + (i+1)\nu] - [x_0 + i\nu]}{[x_0 + I\nu] - x_0} = \frac{1}{I} = \frac{1}{|\Lambda(\mathbb{X};\nu)|} > 0, \tag{A.20}$$

$\forall i \leq I-1$, where $|\Lambda(\mathbb{X};\nu)| = I$ is the cardinality of $\Lambda(\mathbb{X};\nu)$ needed to estimate

[3]This statement will be motivated in the following section.

the probability

$$\mathbb{P}\{x_i < X(\omega) \leq x_{i+1}\} \equiv \mathbb{P}\left\{X(\omega) \doteq l_i\right\} = q, \quad \forall i = 0, \ldots, I-1. \tag{A.21}$$

As it is clear, this is the discrete *uniform* case hypothesis for both the gross-grained partition $\Pi(\mathbb{X}; \nu)$ and the income level set $\Lambda(\mathbb{X}; \nu)$, i.e.

$$X \sim U_d(x_0, x_I) \Leftrightarrow L \sim U_d(l_0, l_{I-1}), \tag{A.22}$$

with probability mass function given by (A.21), a linearly increasing distribution

$$\begin{aligned}\mathbb{P}\{X \leq x_i\} = Q(x_i) &= \frac{x_i - x_0}{x_I - x_0} = \frac{\left(l_i - \frac{\nu}{2}\right) - \left(l_0 - \frac{\nu}{2}\right)}{\left(l_{I-1} + \frac{\nu}{2}\right) - \left(l_0 - \frac{\nu}{2}\right)} \\ &= \frac{l_i - l_0}{l_{I-1} - l_0 + \nu} = Q(l_i) = \mathbb{P}\{L \leq l_i\} = qi > 0,\end{aligned} \tag{A.23}$$

$\forall i = 0, \ldots, I-1$, and, by means of (A.12), expected value

$$\begin{aligned}E(L) &= \sum_{i=0}^{I-1} l_i q(l_i) = qI\left(l_0 + (I-1)\frac{\nu}{2}\right) = \frac{l_{I-1} + l_0}{2} = \bar{L}, \\ E(X) &= \sum_{i=0}^{I} x_i q(x_i) = qI\left(x_0 + I\frac{\nu}{2}\right) = \frac{x_I + x_0}{2} = \bar{X},\end{aligned} \tag{A.24}$$

where

$$\bar{L} = \frac{l_{I-1} + l_0}{2} = \frac{\left(x_I - \frac{\nu}{2}\right) + \left(x_0 + \frac{\nu}{2}\right)}{2} = \frac{x_I + x_0}{2} = \bar{X}. \tag{A.25}$$

In the example discussed here, $x_0 = 10^4\delta$ and $x_I = 10^8\delta$ give $V_{\mathbb{X}} = 10^4\left(10^4 - 1\right)\delta$, $I = 8$ and $\nu = 1,250 \times 10^4\delta$, then $q = 0.1250$. Accordingly, $E(X) = \bar{X} = 5,000.5 \times 10^4\delta$. The problem is that there is no evidence for income to obey the uniform hypothesis because the top incomes are very few, maybe even less than those in absolute poverty, whereas there is a huge bunch of people in the middle.

Assuming $X \sim U_d(x_0; x_I)$ implies that every outcome has the same probability to realize, i.e. events are equiprobable. This may be a starting point in the absence of information, but it nevertheless remains an inappropriate assumption. Even though $\mathbb{X}$ is the only quantitative information available, there may be phenomenological information to consider. For instance, the hypothesis $X \sim U_d(x_0, x_I)$ is inappropriate if evidence shows X is so heterogeneous over the population Ω such that, in samples other than Ω_n, there are realizations which are highly improbable with respect to others. That is, experience tells that if one is going to draw a sample of $n \gg 1$ wage earners from a large population then there is very a small probability to observe a top income. This is a generic regularity and it informs on the phenomenology underlying $\mathbb{X}$. Accordingly, beyond $\mathbb{X}$, the phenomenological information suggests abandoning the uniform hypothesis as unrealistic.

To overcome the unrealistic implications of $X \sim U_d(x_0, x_I)$ it is worth taking a different perspective inspired by information theory.

Realizations $X(\omega)$ of $X : \Omega_n \to \mathbb{X}$, or $X : \Omega_n \to \Pi(\mathbb{X}; \nu)$ whatever ν is, are measurement-events. By sampling Ω_n from Ω one does not know who is going to be drawn at random. Accordingly, by measuring X on $\omega \in \Omega_n$ one does not *ex-ante* know what the realization $X(\omega)$ will be. Nevertheless, the information content of $X(\omega)$ depends on the probability $\mathbb{P}\{X(\omega) \in A_i\} = \mathbb{P}\{X(\omega) \doteq l_i\}$ for $X(\omega)$ to realize in the realization field $\mathbb{X}$, $\Pi(\mathbb{X}; \nu)$ or $\Lambda(\mathbb{X}; \nu)$. The higher the probability for $X(\omega) \in A_i$ or $X(\omega) \doteq l_i$, the lowest the information content. Indeed, a common event is almost a regularity, something one would expect, while a rare event may come unexpected. Said differently, common events like realizations about the expected value are poorly informative compared to those on the longest tail of the distribution under inference.

The amount of information, or self-information, of $X(\omega)$ to realize l_i depends on the probability $\mathbb{P}\{X(\omega) \doteq l_i\}$ for $l_i \in \Lambda(\mathbb{X}; \nu)$ and it is indicated as $\mathcal{I}(X(\omega)) = -\log_a[X(\omega)]$: if $a = 2$, then $\mathcal{I}(X(\omega))$ is measured in *bits*; if $a = 10$, it is measured in the standard unit *hart*; if $a = e$, the unit of measure is the *nat*, which is that chosen for the following.

One may then consider a second hypothesis: since, at least beyond a certain threshold, the higher the income the fewer the wage earners, it then seems reasonable to consider an exponentially decaying distribution to represent such a phenomenology. From $X \sim U_d(x_0, x_I)$ one defines $L \sim U_d(l_0, l_{I-1})$ as in (A.22): both follow the same distribution (A.23). Define then a new random variable by means of the following non-linear transformation

$$Y = -\frac{\ln(L)}{\theta}, \quad \theta > 0, \tag{A.26}$$

replacing L with its self-information and rescaling it on the parameter θ. Being $\ln(\cdot)$ monotone it preserves the ordering but the minus sign reverses the order. This introduces the previous phenomenological information: the almost improbable realizations receive now a higher attention, contrary to those which are almost regular. Since $L \sim U_d(l_0, l_{I-1})$ with $q(l_i) = q$, then the probability function of Y is

$$\begin{aligned} \mathbb{P}\{Y \le y_i\} &= \mathbb{P}\left\{-\frac{\ln(L)}{\theta} \le y_i\right\} = \mathbb{P}\{\ln(L) \ge -\theta y_i\} \\ &= \mathbb{P}\{L \ge e^{-\theta y_i}\} = 1 - \mathbb{P}\{L < e^{-\theta y_i}\} \\ &= 1 - q e^{-\theta y_i} = Q_Y(y_i), \end{aligned} \tag{A.27}$$

which implies

$$q_Y(y_i) = q\theta e^{-\theta y_i} = q\theta l_i, \tag{A.28}$$

where e denotes $\exp(1)$.

This result shows that the probability function of transformed measurement-events on $\Lambda(\mathbb{X}; \nu)$ non-linearly decreases as an exponential random variable at the rate θ.

This is precisely what expected: extreme values toward the upper bound of $\mathbb{X}$ should be almost improbable as top incomes are very few.

But where does θ come from? Since $q_Y(y_i)$ is a probability mass function, then

$$\begin{aligned}\sum_{i=0}^{I-1} q_Y(y_i) = 1 \Rightarrow \sum_i q\theta l_i = 1 \Rightarrow q\theta \sum_i (l_0 + i\nu) = 1 \\ \Rightarrow q\theta\left(Il_0 + \nu\sum_i i\right) = 1 \\ \Rightarrow q\theta\left[Il_0 + I(I-1)\frac{\nu}{2}\right] = 1 \\ \Rightarrow q\theta I\left[l_0 + (I-1)\frac{\nu}{2}\right] = 1 \\ \Rightarrow \theta E(L) = 1.\end{aligned} \tag{A.29}$$

Therefore, according to (A.24) and (A.25) it follows that

$$\theta = \frac{1}{E(L)} = \frac{1}{\bar{L}}, \quad \bar{L} = \bar{X}. \tag{A.30}$$

Hence, the decay parameter θ of the discrete-exponential is the reciprocal of the expected value of the discrete-uniform hypothesis. The higher the expected value of income under the uniform hypothesis—that is, the wider is the realization field $\mathbb{X}$—the lower the decay rate θ regulating the self-information Y of L. Interestingly enough, one then obtains the expected value estimator of Y as

$$\begin{aligned}E(Y) &= \sum_i y_i q_Y(y_i) = -\theta\sum_i \frac{\ln(l_i)}{\theta} l_i = -\sum_i l_i \ln(l_i) \\ &= S[\Lambda(\mathbb{X};\nu)],\end{aligned} \tag{A.31}$$

which is the Shannon entropy of the set $\Lambda(\mathbb{X};\nu)$.

Few remarks are now worth stressing. Values $l_i \in \Lambda(\mathbb{X};\nu)$ depend on $x_i \in \Pi(\mathbb{X};\nu)$ and $\nu = \nu(V_{\mathbb{X}}; I) = V_{\mathbb{X}}/I$, being $\mathbb{X}$ endowed with the ERUM δ. Therefore

$$L \sim U_d(l_0, l_{I-1}) \Rightarrow Y = -\frac{\ln(L)}{\theta} \sim \mathcal{E}(\theta), \tag{A.32}$$

where $\theta = \frac{1}{E(L)}$ and $E(Y) = S[\Lambda(\mathbb{X};\nu)]$. Under the uniform hypothesis at orders of magnitude higher than the ERUM nothing will change beyond numerical aspects. Indeed, starting from $\pi(\mathbb{X};\delta)$ in (A.9) one may adopt a definition of ν different from (A.12) to end up with a different $\Lambda(\mathbb{X};\nu)$ of (A.17) but, as long as this is the origin of the uniform hypothesis, nothing will change.

Finally, if the probability field $\mathbb{I} = [0; 1]$ of the observable X is isotropic with respect to the homogeneous partition $\Pi(\mathbb{X};\nu)$ of the realization field $\mathbb{X}$, then the observable L on $\Lambda(\mathbb{X};\nu)$ follows a uniform hypothesis which can be reconciled with a discrete exponential observable Y, whose expected value is the Shannon entropy of the measurement-event set $\Lambda\ (\mathbb{X}; \nu)$.

All these results are interesting but, unfortunately, neither the uniform nor the exponential hypotheses seem to be appropriate to describe the income distribution. The uniform model may seem appropriate on a restricted portion of outcomes, where the distribution is almost linear as $Q\ (l_i)$, while the exponential one may seem appropriate on that portion of the realization field where the probability to sample sufficiently large incomes is very low and decays as $Q_Y\ (y_i)$. None of the two satisfactorily covers the whole range of incomes and both do not describe the left-fat and right-long tails. So, which story do these models tell about the inference of the distribution of a given transferable quantity endowed with an ERUM δ and defined on a limited and closed set $\mathbb{X} = [x_{\min}; x_{\max}]$? Why is it so hard making inference upon them? Roughly: what is the problem with monetary and transferable quantities? The problem is that many transferable quantities characterized by an ERUM behave *strangely*, i.e. *stochastically* and (in a sense) *relativistically* at the same time, and so does income and wealth. Therefore, no *transformation* operating in the same way and everywhere on $\mathbb{X}$ allows for a unified model to describe the whole distribution. This is because in the case of strange quantities the relevance of realizations' information content is not the same everywhere, which means the probability field is not isotropic w.r.t. homogeneous partitions of the realization field.

To understand and model the *strange mechanics of monetary and transferable quantities* it needs introducing a more sophisticated tool than transformation: the *deformation* on the realization field $\mathbb{X}$ and a suitable deformed mathematics of functions operating on it. This is because it is as if single masses $X\ (\omega)$ on the realization field were able to interact with the probability field, that is deforming its shape. There is no transformation, neither monotone nor non-linear, which is capable to capture such a kind of interaction because transformation outcomes preserve the topological properties of the space they apply to. Stated otherwise, if $\mathbb{X} \subset \mathbb{R}$ then every regular transformation $\mathcal{T} : \mathbb{X} \to \mathbb{Y}$, with $\mathbb{Y} \subset \mathbb{R}$, is preserving the Euclidean topological properties, which is not what required to grasp the strange behavior of strange quantities. As anticipated in Sect. A.1, it needs something like the Lorentz transformation, which reconciles the Galilean relativity on the Euclidean space through time with the Einstein interpretation of relativity principles on the non-Euclidean Minkowski space-time. We will see that by introducing an appropriate deformation, and a suitable deformed mathematics, the κ-generalized model arises almost naturally to capture the phenomenology of strange quantities.

A.5 The *Classic* MaxEnt Inference Method

In previous sections a weak analogy between the ERUM δ on $\mathbb{X}$ and the real infinitesimal $\mathrm{d}x$ on a proper subset of $\mathbb{R}$ has been discussed. While being aware that, by nature, the support $\mathbb{X}$ of the transferable quantity X with ERUM δ is discrete, it has been nevertheless accepted the idea of using the infinitesimal calculus formalism if the realization field is assumed to be sufficiently dense and that its width is very large enough compared to the ERUM or to a conventionally chosen CRUM.

Let $f(\cdot;\boldsymbol{\theta}) : \mathbb{X} = [x_{\min}; x_{\max}] \to \mathbb{I} = [a; b]$ be a function modulated by a set of parameters $\boldsymbol{\theta} \in \boldsymbol{\Theta} \subset \mathbb{R}^d$ such that $a < b < \infty$. $f(\cdot;\boldsymbol{\theta})$ is assumed (at least right-) continuous and integrable on $\mathbb{X}$: right-continuity can be always accepted by specifying f in such a way that $f(u;\boldsymbol{\theta}) = f(x_h;\boldsymbol{\theta})$ for any $u \in (x_h; x_{h+1}]$, and the more the partition of the realization field $\mathbb{X}$ is fine-grained the more this is consistent. Being $f(x_h;\boldsymbol{\theta})$ integrable on $\mathbb{X}$, then $\int_{x_{\min}}^{u} f(x_h;\boldsymbol{\theta})\,\mathrm{d}x = F(u;\boldsymbol{\theta}) - F(x_{\min};\boldsymbol{\theta})$. Clearly, $F(\cdot;\boldsymbol{\theta})$ is increasing (strictly if f is continuous, or by steps if f is right-continuous) and differentiable: $\frac{\mathrm{d}F(u;\boldsymbol{\theta})}{\mathrm{d}x}|_{(x_h;x_{h+1}]} = f(x_h;\boldsymbol{\theta})$. Moreover, $F(x_{\min};\boldsymbol{\theta}) < F(x_{\max};\boldsymbol{\theta}) < \infty$.

If all such assumptions are plausible, one may then define the Shannon entropy

$$S(X; f) := \int_{\mathbb{X}} \mathcal{I}(f(x;\boldsymbol{\theta}))\, f(x;\boldsymbol{\theta})\,\mathrm{d}x = -\int_{\mathbb{X}} f(x;\boldsymbol{\theta}) \ln f(x;\boldsymbol{\theta})\,\mathrm{d}x. \quad \text{(A.33)}$$

Being $\mathcal{I}(f(x;\boldsymbol{\theta}))$ continuous, or at least right-continuous, and $\mathbb{X}$ limited and closed, if not compact, then $S(X; f)$ admits an absolute maximum value.

The inferential asks whether is there any privileged functional form for $f(x;\boldsymbol{\theta})$ to maximize the entropy $S(X; f)$ under the constraint

$$\int_{\mathbb{X}} f(x;\boldsymbol{\theta})\,\mathrm{d}x = \alpha. \quad \text{(A.34)}$$

Answering this question needs a constrained optimization problem to be solved by means of Lagrange multipliers. For the moment the integration domain $\mathbb{X}$ can be neglected and recalled at the end to find a closed-form solution. The Lagrangian is

$$\mathcal{L}(f, \lambda_1) = -\int f(x;\boldsymbol{\theta}) \ln f(x;\boldsymbol{\theta})\,\mathrm{d}x - \lambda_0 \left[\int f(x;\boldsymbol{\theta})\,\mathrm{d}x - \alpha\right]. \quad \text{(A.35)}$$

The first order condition w.r.t. the unknown f gives

$$\frac{\partial \mathcal{L}}{\partial f} = -x \ln f(x;\boldsymbol{\theta}) - (1+\lambda_0)\,x + u = 0, \quad \text{(A.36)}$$

where u is an arbitrary integration constant. Solving for f yields

$$f(x;\boldsymbol{\theta}) = e^{-(1+\lambda_0)} + u \quad \text{(A.37)}$$

as candidate to solve the problem. To this end, (A.37) must satisfy the constraint (A.34), i.e.

$$\int_{\mathbb{X}} f(x;\boldsymbol{\theta})\,\mathrm{d}x = e^{-(1+\lambda_0)} \int_{\mathbb{X}} \mathrm{d}x + u \int_{\mathbb{X}} \mathrm{d}x = \alpha, \quad \text{(A.38)}$$

which gives

$$\alpha = e^{-(1+\lambda_0)} V_{\mathbb{X}} + u V_{\mathbb{X}}, \quad V_{\mathbb{X}} = x_{\max} - x_{\min}. \tag{A.39}$$

Therefore

$$u = \frac{\alpha}{V_{\mathbb{X}}} - e^{-(1+\lambda_0)} \tag{A.40}$$

yields

$$f(x;\boldsymbol{\theta}) = \frac{\alpha}{V_{\mathbb{X}}} > 0 \Rightarrow \int_{x_{\min}}^{x} f(u;\boldsymbol{\theta})\, \mathrm{d}u = F(x;\boldsymbol{\theta}) = \alpha \frac{x - x_{\min}}{V_{\mathbb{X}}} \geq 0. \tag{A.41}$$

This is the exact solution: $f(x;\boldsymbol{\theta})$ does not depend on x and there is no trace of the parameters $\boldsymbol{\theta}$. Notice that if one sets $\alpha = 1$ in the constraint (A.34) then $F(x_{\max};\boldsymbol{\theta}) = 1$ and $F(x_{\min};\boldsymbol{\theta}) = 0$, hence he is setting a strong condition that allows $f(x;\boldsymbol{\theta})$ being the *uniform density*—and this is why f does not depend on x.

Therefore, maximizing the Shannon entropy (A.33) under the normalization constraint (A.34) with $\alpha = 1$ to infer the functional form of the distribution $F(x;\boldsymbol{\theta})$ by knowing only $\mathbb{X} = [x_{\min}; x_{\max}]$ gives the uniform distribution. In the previous section the uniform distribution was assumed; now such a hypothesis receives an inferential interpretation: in absence of any further information beyond the realization field $\mathbb{X}$, the best one can do at inferring the distribution of the random variable X is to associate the same density to all the outcomes.

What is the meaning of this procedure? Specifying the problem

$$\begin{cases} \max_f & S(X;f) := -\int_{\mathbb{X}} f(x;\boldsymbol{\theta}) \ln f(x;\boldsymbol{\theta})\, \mathrm{d}x \\ \text{s.t.} & \int_{\mathbb{X}} f(x;\boldsymbol{\theta})\, \mathrm{d}x = 1 \end{cases} \tag{A.42}$$

sets an inferential problem based on the unique information available, i.e. the knowledge of $\mathbb{X} = [x_{\min}; x_{\max}]$. The inferential question is not an easy one and the available information is poor: therefore, the answer cannot be more satisfactory than the uniform solution. The differences with respect to the results of the previous section are that (i) the random variable X and its support $\mathbb{X}$ are *here* assumed to be continuous and that (ii) the uniform solution comes through a constrained optimality criterion, whereas *there* the random variable was assumed to be discrete-uniform on its homogeneous discrete support. The main difference, however, is not in discreteness and continuity—indeed the maximum entropy (MaxEnt) problem could be specified in both cases—rather it is that the MaxEnt solution gives an inferential interpretation to what has been previously assumed for convenience. Nevertheless, due to its implications, the simplest MaxEnt problem returns a uniform distribution which is not suitable for inference on the income distribution: in fact, it is very unlikely assigning to income-receiving units a linearly increasing probability $\mathbb{P}\{X \leq x\} = F(x)$ when evidence shows that the top incomes are fewer than absolutely poor ones and there is a huge bunch on people in the middle range. Hence, the simplest MaxEnt problem is nothing but an introduction to more sophisticated inferential problems.

Two ways can be basically isolated to proceed with the issue of inference. The first method is empirically based to meet an analytic interpreter of the data. Once data have been appropriately structured, by means of transformations and classifications suitable for the phenomenon under scrutiny, a histogram or empirical density is obtained. After that, an analytic probabilistic model is chosen among many others as the theoretical density ruled by some parameters to be estimated on sample data. Finally, a goodness-of-fit test is adopted to decide whether the analytic model is a good interpreter of the empirical density: if the test does not refuse the null hypothesis, then the analytic model is assumed as a good interpreter of the sample data generating process, at least *until proved otherwise*. This method is almost standard and often works pretty well, but it hides two problems: (i) one can opportunely prepare data in order to meet an analytic model not to be refused by the goodness-of-fit test; (ii) there can be more than a single model to be eligible as good interpreters of the data-generating process, in which case one may decide upon efficiency and parsimony criteria subject to desirable properties related to the phenomenon of interest. Therefore, in general, there is not absolute certainty upon the chosen model but only a confidence due to some statistical significance criteria.

Although still evidence-based, the second approach is different as is much more extensive by involving information about the system of observations as a whole together with individual sampling data. By reasoning about the phenomenology of events the density must represent—i.e. by offering a description based on the behavior of the quantity and of the units of observation—and involving some optimality criterion characterizing the theoretical model to be used, some constraints (either numeric or functional) are set to infer the functional form that, consistently with all available information, allows for inference with minimum uncertainty of density which is more likely to be eligible at interpreting the data-generating process of sample data—not until otherwise proved, but consistently with known information. In this case transformations and classifications have a minor relevance, whereas two other aspects are important: (i) the choice of the objective function to be optimized and (ii) the specification of optimization constraints which embed in the model the known phenomenological aspects.

Both approaches are empirically based and make use of sampling data, but the underlying philosophy of the two is different. In the first approach, by means of parsimony and statistical significance criteria, the choice is made among a collection of *ex-ante* feasible models. In the second approach data tell all what they have to tell for identifying the less improbable model: the choice is made *ex-post*, consistently with both sampling and systemic data, and it may also happen that the results were unknown before, i.e. new probabilistic models may be discovered to accommodate with data.

Not depending on the chosen approach, one ends up with a probabilistic model for the random variable X with density $f(x;\boldsymbol{\theta})$ characterized by a set of parameters $\boldsymbol{\theta} = (\theta_1, \ldots, \theta_d)$. As $\boldsymbol{\theta}$ changes on a parametric space $\boldsymbol{\Theta}$, the function $f(x;\boldsymbol{\theta})$ determines a specific member of a given family of distributions. The family of distributions whose members have a density of the following form

$$f(x;\boldsymbol{\theta}) = D(x)\, e^{\sum_j A_j(\boldsymbol{\theta})B_j(x) - C(\boldsymbol{\theta})}, \quad D(x) \geq 0, \tag{A.43}$$

is called the *exponential family*. Members of this family are the binomial, Poisson, Weibull, normal, lognormal and many others—of course the exponential distribution belongs to such a family, but surely the uniform does not, although, as shown before, the uniform generates the exponential distribution.

As for the problem (A.42), assume having an additional information: the estimate of the average of X, i.e. $\bar{X} = m$, which is a quantitative datum about the system as whole, loosely informing about the overall phenomenology. The problem becomes

$$\begin{cases} \max_f & S(X; f) := -\int_{\mathbb{X}} f(x; \boldsymbol{\theta}) \ln f(x; \boldsymbol{\theta})\, \mathrm{d}x \\ \text{s.t.} & \int_{\mathbb{X}} f(x; \boldsymbol{\theta})\, \mathrm{d}x = 1 \\ & \int_{\mathbb{X}} x f(x; \boldsymbol{\theta})\, \mathrm{d}x = m \end{cases} \tag{A.44}$$

and is just a little bit more complicated to be solved. The Lagrangian is

$$\begin{aligned} \mathcal{L}(f, \lambda_0, \lambda_1) =& S(f) - \lambda_0 \left[\int_{\mathbb{X}} f(x; \boldsymbol{\theta})\, \mathrm{d}x - 1 \right] \\ & - \lambda_1 \left[\int_{\mathbb{X}} x f(x; \boldsymbol{\theta})\, \mathrm{d}x - m \right]. \end{aligned} \tag{A.45}$$

Without developing all the mathematical steps, the first order condition w.r.t. the unknown f gives

$$f(x; \boldsymbol{\theta}) = e^u e^{-\frac{\lambda_1}{2} x - (1+\lambda_0)} \tag{A.46}$$

as the suitable candidate to solve the problem. Such a proposal depends on the multipliers $\boldsymbol{\lambda} = (\lambda_0, \lambda_1)$ associated to the constraints and on the integration constant u which is due to the fact that the support $\mathbb{X}$ is limited and closed between $x_{\min}$ and $x_{\max}$. Such multipliers can be specified as shown in Chap. 3, but it can be anyway observed that they depend on the constraints which allow for introducing the additional information about the system.

Not depending on the estimated values of the unknown parameters, a general result is not missed. Having imposed a set of constraints as selection criteria under optimality condition of the choice, parameters of the density to be inferred depend on the multipliers and *vice versa*: $\boldsymbol{\theta} = \theta(\boldsymbol{\lambda})$. Therefore, without loss of generality, one can set

$$\begin{cases} A_1 [\theta(\boldsymbol{\lambda})] = 0 \\ A_2 [\theta(\boldsymbol{\lambda})] = -\dfrac{\lambda_1}{2} \\ B_1(x) = \text{any value} \\ B_2(x) = x \\ C[\theta(\boldsymbol{\lambda})] = 1 + \lambda_0 \\ D(x) = e^u \end{cases} \tag{A.47}$$

so that (A.46) may be written as

$$f\left[x;\theta\left(\boldsymbol{\lambda}\right)\right]=e^{u}e^{\{A_1[\theta(\boldsymbol{\lambda})]B_1(x)+A_2[\theta(\boldsymbol{\lambda})]B_2(x)-C[\theta(\boldsymbol{\lambda})]\}},\tag{A.48}$$

which proves that the inferred density belongs to the exponential family.

To gain a sounder idea of the solution one can search out the estimator of the integration constant as a function of the multipliers, i.e. of the parameters. To this end, it is sufficient substituting (A.46) in the normalization constraint to end up with

$$\begin{aligned} e^{u} &= \frac{\lambda_1}{2}e^{(1+\lambda_0)}\left[e^{-\frac{\lambda_1}{2}x_{\max}}-e^{-\frac{\lambda_1}{2}x_{\min}}\right]^{-1}>0 \\ &\Rightarrow u:=u\left(\lambda_0,\lambda_1,x_{\min},x_{\max}\right), \end{aligned}\tag{A.49}$$

which is always fulfilled $\forall\lambda_1>0$ because $x_{\max}\gg x_{\min}$. Once the multipliers have been determined, an estimate of the integration constant will become explicit, but for the moment, this is not relevant. Therefore, set

$$e^{u}=e^{u(\lambda_0,\lambda_1,x_{\min},x_{\max})}\equiv D>0.\tag{A.50}$$

Hence the density candidate to solve the problem becomes

$$f\left[x;\theta\left(\boldsymbol{\lambda}\right)\right]=De^{-(1+\lambda_1)}e^{-\frac{\lambda_1}{2}x},\tag{A.51}$$

which at this point surely satisfies the normalization constraint. Therefore

$$\int_{\mathbb{X}}f\left[x;\theta\left(\boldsymbol{\lambda}\right)\right]=De^{-(1+\lambda_1)}\int_{\mathbb{X}}e^{-\frac{\lambda_1}{2}x}=1\tag{A.52}$$

in such a way that

$$e^{-(1+\lambda_0)}=\frac{1}{D\int_{\mathbb{X}}e^{-\frac{\lambda_1}{2}x}}\equiv Z\left(D,\lambda_0,\lambda_1\right),\tag{A.53}$$

where $Z\left(D,\lambda_0,\lambda_1\right)$ is the so called *partition function*. Finally

$$f\left[x;\theta\left(\boldsymbol{\lambda}\right)\right]=\frac{e^{-\frac{\lambda_1}{2}x}}{Z\left(D,\lambda_0,\lambda_1\right)}\tag{A.54}$$

is a member of the exponential family distributions and assumes the *Gibbs' form*: once the multipliers will be made explicit, such a solution will receive an appropriate parametrization and it will not change its species.

Therefore, entropy maximization under constraints on moments—even higher than the first—by setting numeric values always induces a density in the exponential family reducible to a Gibbs' form: if such constraints are not numeric but functional, i.e. introducing the desired phenomenology, the solution may also belong to a different family of distributions, as is happens in the case of the κ-generalized distribution—see Chap. 3.

It is now worth noticing that the exponential model has been inferred by means of the same procedure involved to infer the uniform distribution, with the only difference that one more constraint about the average was added. In Sect. A.4, after having assumed the uniform hypothesis, the exponential model was obtained by means of a logarithmic transformation of the uniform model rescaled on the reciprocal of uniform expected value estimator. Both approaches lead to the same result, but the one developed here provides an inferential interpretation of that assumption.

The exponential result with several constraints is more sophisticated than the uniform one but it is still not appropriate and it needs knowing estimates for the moments $E\left(X^j\right) = m_j$. As a matter of fact, this is the simplest representation; if instead of involving numerical estimates of moments one would fix their estimators, i.e. functions, then the results would generalize.

However, at this point a *logical circularity* problem arises: to determine the moment estimators one needs to know the density while, here, the density is needed to set moment estimates, for which estimators are required. Therefore, it seems there is no easy way out from this *chicken-and-egg problem*. The reader will surely remember that there are at least to approaches to deal with inference: (i) goodness-of-fit based inference, which is what can be done by fixing numerical values for moments, and (ii) inference from the underlying phenomenology, which is what can be done by fixing the estimators—which means opportunely specifying the constraints of the MaxEnt problem. This second perspective gives proof of the inferential power of the MaxEnt methodology. For instance, consider the following simple problem

$$\begin{cases} \max_f & S(f) := -\int_{\mathbb{X}} f(x;\boldsymbol{\theta}) \ln f(x;\boldsymbol{\theta})\,\mathrm{d}x \\ \text{s.t.} & \int_{\mathbb{X}} f(x;\boldsymbol{\theta})\,\mathrm{d}x = 1 \\ & \int_{\mathbb{X}} g_1(x;\boldsymbol{\gamma}) f(x;\boldsymbol{\theta})\,\mathrm{d}x = \mu_1(\boldsymbol{\psi}) \\ & \dots \\ & \int_{\mathbb{X}} g_J(x;\boldsymbol{\gamma}) f(x;\boldsymbol{\theta})\,\mathrm{d}x = \mu_J(\boldsymbol{\psi}) \end{cases} \tag{A.55}$$

where $g_j(x;\boldsymbol{\gamma})$ is a function obeying the required phenomenology of the random variable X and $\mu_j(\boldsymbol{\psi})$ is the estimator of the corresponding expected value. Under these conditions, solutions out of the exponential family may emerge. The more the result is a good interpreter of the data, the more one is satisfied both empirically and theoretically. In the case the outcome belongs to the exponential family the result would not be too satisfactory because income or wealth exhibits an exponential decay only beyond a given threshold, where it is better explained by a power-law family distribution with long right tail. Therefore, even the more complex exponential model is not really appropriated to interpret the income or wealth data generating process. Singh and Maddala (1976) explain this aspect in terms of increasing (exponential) and decreasing (power-law) failure rate of a distribution, whereas the perspective followed here is rather different: the uniform and exponential models are intimately entangled and both are suitable only if the realizations of the random variable are dominated by some regularity, proportionality or homogeneity principles over the whole support.

A.6 Phenomenology of *Strange* Quantities

The characteristics of *strange quantities* have only been introduced in Sect. A.1 and have been implicitly involved up to this point. Such properties will be now discussed in more detail to summarize the phenomenology of strange quantities: this should help understanding the need to involve the *mathematics of deformations*. Moreover, in addition to what has been already anticipated, it can also be specified that strange quantities are characterized by a micro-funding principle of their aggregate outcomes, which can be reconciled with the *kinetic interaction principle* (KIP) introduced by Kaniadakis (2001) to develop a relativistic interpretation of the statistical mechanics while developing a complete mathematics of deformations—see Kaniadakis (2002, 2005).

To provide an interpretation of the physics introduced by Kaniadakis requires justifying the mathematics of deformations, needed to model the phenomenology of strange quantities. Also, the logic transition from a *world of regular quantities* to a *world of strange quantities* requires an interpretation based on the analogy with ontology: this allows taking a step beyond the formal technique in order to approach a methodology to develop theoretical proposals. Indeed, both in economics and physics the ontology of complex systems allows to make clear what are the observation units, what is meant to be observed and how it is observed. Moreover, analogy allows for explaining how observation units behave, what are the mechanisms generating emergent macroscopic phenomena, and what are the techniques suitable for inference. A detailed development of both aspects is far beyond the aim of this section: in this regard, Sects. A.1–A.3 may furnish a sufficient introduction. Nevertheless, to trace an analogy between some aspects of physics and economics in the present field of investigation—i.e. in the context of income and wealth distributions—a synthesis is proposed below.

As already anticipated, a strange quantity melts stochasticity and relativism. On the one hand, it is intelligible with a probabilistic interpretation; on the other hand, one should consider that the realization field is not homogeneous: i.e. different statistical principles dominate different portions of the support depending on the orders of magnitude of feasible realizations—see Sect. A.3. Therefore, the probability field is not isotropic with respect to partitions of the realization field, even after numerical transformations of data—see Sects. A.4 and A.5.

The relativistic character of a *strange* quantity X is due to its strange behavior obeying different statistical principles on different portions of the support $\mathbb{X} = [x_{\min}; x_{\max}]$, that is depending on the orders of magnitude of feasible realizations for measurement-events: to avoid involving an inopportune terminology in the economic field, a weak analogy with physics is preferred by saying that a strange quantity is a-symmetric. In the simplest case this a-symmetric character means that for a given threshold ξ the exponential family probabilistic principle rules the stochastic behavior below this value, i.e. $\forall x \leq \xi : X \sim \mathcal{E}(\boldsymbol{\theta})$, while beyond the threshold the power-law family probabilistic principle characterizes the realizations at high order of magnitude, i.e. $\forall x > \xi : X \sim \mathcal{P}(\boldsymbol{\theta})$. Therefore, the probabilistic principle governing a strange quantity seems not to be unique over the whole support but *relative* to subsets with different orders of magnitude of realizations.

In the case of income and wealth, such an a-symmetry (i.e. relativism) means these quantities satisfy something like the weak Pareto principle while apparently following models like the Pareto, Dagum type I or Singh-Maddala. In addition, it often happens that models which are good interpreters of the distribution right tail are not so also for the left one. Therefore, in such case one prefers estimating a point at which the distribution changes to describe the first part according to an exponential model and the second with a power-law one. On the contrary, as it will be shown, the κ-generalized distribution embeds both models so as to reconcile the two probabilistic principles into a new family of distributions.

These aspects were already introduced in Sect. A.1. To explain their genesis and implications a theoretical justification can be developed by resorting to categories of complex system theory, since economies are complex systems and, as such, statistical mechanics operates as the appropriate descriptive and inferential methodology. Moreover, being the κ-generalized model obtained in the frame of the relativistic interpretation of statistical mechanics, it becomes clear that it is the appropriate model for strange quantities.

Three different kinds of a-symmetry can be isolated. *Structural a-symmetry* is related to heterogeneity of system constituents at different levels of granularity between the very micro and the very macro levels. *Behavioral a-symmetry* is related to interaction of system parts and reads in terms of transferability of portions or slots of a quantity between two or more system constituents (direct interaction) and between two or more sub-systems (indirect interaction). *Shape a-symmetry* is related to the asymmetric shape of the density representing the system as a whole, and it is also related to two emergent phenomena which are due to the continuous interactions of heterogeneous constituents: concentration and inequality.[4]

Finally, but this is to be discussed at first, there is another important aspect characterizing the phenomenology of strange and regular economic quantities: they are intrinsically quantized or discrete, not conservative because they are subject to growth *stimuli*, and elastically-bounded, which means their support is always limited and closed even if the bandwidth may change through time.

[4]Contrary to the standard language, these terms are not used here as synonymous (see Gallegati et al. 2015).

A.6.1 Discreteness, Growth and Elastic Bandwidth Dynamics

As anticipated in Sect. A.5, transferable quantities with ERUM $\delta = 0.01\mathbb{D}$—even not monetary—are discrete by nature. Therefore, their changes happen due to transfers of slots of ERUM among observation units: in analogy with physics speech, it could be said that they are intrinsically quantized even at high orders of magnitude.

Biologic organisms, human beings and social systems are all subject to the *stimuli* of growth which, very roughly, can be seen as the latent principle driving the evolution of a living entity with the aim of improving its present state with respect to certain functions or objectives toward some degree of maturation or satisfaction fulfillment. However, intuitively enough, nothing can grow indefinitely through time and nothing can grow unlimited, and this is almost evident in the case of socio-economic systems: resources, both material and immaterial, are not indefinite, and monetary ones are not too. Monetary resources are conventional institutions facilitating transactions: in this sense they may be units of some money paper in a given currency, but also other materials used for this purpose. Anyhow, without going into deep with these more philosophical aspects, it is clear that monetary resources, as well as income and wealth, may grow faster or slower through time but not with unlimited speed and not incommensurately.

Considering no debt, income and wealth (both indicated as X) have a clear absolute lower bound, that is $x_{\min} = 0$, but there is not a precise value limit for the upper bound other than the income or wealth of the most rich person on hearth at a given time. Moreover, when dealing with sample data, such limits may change from sample to sample, from economy to economy, from period to period. However, a maximum value for income and wealth exists in the population where the samples have been drawn from, so as at the time of Egyptian Pharaohs as well as at the present day in advanced capitalistic economies. The highest fortune one may observe nowadays is that of Bill Gates, which has been estimated in 2014 at about 76 billions of dollars, with an annual income of 11.5 billions of dollars.[5] Therefore, if an upper bound $x_{\max}$ is to be assumed, then one could consider these values for wealth and income, respectively. However, even though such an upper bound exists, it could not always be the case that it remains constant through time and space; therefore, one should accept the idea that the upper bound may dynamically change, also due to growth or, at least, in connection with growth. Hence, mainly dealing with panel sample data, the *bandwidth* of income and wealth dynamically change in the long run, but always with a finite limit: that is $X(\omega, t) \in \mathbb{X} = [x_{\min}; x_{\max}]$ and $0 \leq x_{\min} \ll x_{\max} < +\infty$.

Thus, no quantity relevant to economics is unbounded, but they are all finite and limited: even under the growth principle one should at least consider the average, as done in Sect. A.5 with regard to the baseline MaxEnt inferential model. Indeed, the meaning of the expected value in a specific constraint of the MaxEnt problem is also to involve a limit for the state variable X to make the inference consistent with

[5] See http://www.forbes.com/profile/bill-gates/?list=billionaires and http://www.paywizard.org/main/salary/vip-check/bill-gates.

an observed aggregate datum: the average is nothing but the aggregate value of X divided by the number of observation units.

However, one remembers that maximizing the Shannon entropy under constraints on moments leads to a distribution which is a member of the exponential family, such as a the Maxwell-Boltzmann distribution in the form of Gibbs. This model is very effective but not strange because, being classically specified, it can not grasp the implications of heterogeneity, interaction, concentration and inequality which are at the roots of the strange behavior of limited and transferable quantities endowed with an ERUM. It is on these aspects that one should intervene by properly deforming the classic model into a strange one to end up with the κ-generalized.

A.6.2 Structural A-Symmetry: Heterogeneity and Granularity

The sample Ω_n is an economic system of n agents ω, heterogeneous w.r.t. endowments $X(\omega)$ of income or wealth X. As discussed in Sect. A.2, the realization field $\mathbb{X}$ has a range $V_{\mathbb{X}}$ very wide, typically of the order of $I = 10^8$ *bins* at the level of the ERUM, variously representable with partitions at different degrees of granularity. Moreover, experience shows that the major part of the realization field is concentrated around the expected value $X(\omega) < \left|\bar{X} \pm \phi\right|$ with a density that at the extrema of this conventional interval performs almost in the same way, i.e. $p\left(\bar{X} - \phi\right) \approx p\left(\bar{X} + \phi\right)$, with a maximum at $\bar{X}$. On this conventional interval of realizations one may find *moderately* poor incomes if $x \in M_- = \left[\bar{X} - \phi; \bar{X}\right]$ and *moderately* rich ones if $x \in M_+ = \left[\bar{X}; \bar{X} - \phi\right]$, hence a *medium class* is conventionally indicated as $M = M_- \cup M_+$.

Out of this interval, the amplitude $\left(\bar{X} - \phi\right) - x_{\min}$ is very much lower than $x_{\max} - \left(\bar{X} + \phi\right)$. Conventionally, set the poor class as $P = \left[x_{\min}; \bar{X} - \phi\right]$ and let $R = \left[\bar{X} + \phi; x_{\max}\right]$ be the rich class. Experience shows $p(x) < p\left(\bar{X} - \phi\right)$ if $x \in P$ and $p(x) < p\left(\bar{X} + \phi\right)$ if $x \in R$; that is: (very) poor and (very) rich incomes are not widespread w.r.t. medium incomes; the probability of sampling very poor and very rich incomes is pretty low. Moreover, there is a little but significant difference between P and R: it can be regularly observed that on P the density decays faster and faster as far as incomes leave the average toward the minimum—this zone covers a restricted set of values—while on R the decay is lower and lower as far as incomes leave the average toward the maximum—this zone covers a wide range of income values. In other words, moving on P from $x_{\min}$ the density rapidly increases, then it slows down around the maximum at the average and finally decreases toward the maximum value of income while decelerating on R as incomes approach x_{max}. Moreover, at some critical level the density changes its concavity and, beyond such a level, the rate of change becomes almost null so that the density becomes flat for a wide range of income values till $x_{\max}$. Therefore, on $P \cup M$ one may assume a density profile typical of the *exponential family of distributions*, while one R one may assume a density profile typical of the *power-law distribution family*.

From these phenomenological observations it is agreed that the density is *asymmetric*, that is mostly concentrated on a wide medium class, loosely concentrated on a restricted poor class and even less concentrated on a very wide rich income class. Stated otherwise, there are lots of medium incomes, few poor incomes within a restricted range, and even fewer rich ones within the widest range. Also, within each of these conventional macro-classes there are several degrees but $X(\Omega_n)$ does not realize all the feasible values in the finest grained partition of $\mathbb{X}$: states on the support X are more densely populated about the center, less toward the minimum, and much less toward the maximum income. Indeed, as discusses in Sect. A.3, it needs a huge number of poor people $\omega_0 \in P$ to balance with a single rich $\omega_1 \in R$. Therefore, a first element is that *heterogeneity* is really extreme, it really matters and it is persistent at different degrees of refinements of the partitions of the realization field $\mathbb{X}$.

A.6.3 Behavioral A-Simmetry: Interaction and Transferability

A second trait is that transferable quantities are so because they allow for withdrawn of an amount of income or wealth from one unit to transfer it to another unit, even without changing the aggregate value.[6]

Let ω_R be a rich and ω_P a poor: transfers $\omega_P \leftrightarrow \omega_R$ are interactions in the form of flows or transactions. If at least a ω between the two is an individual, either an elementary observation unit or collective, then interaction (e.g. a good by-sell relation, tax payment, inter-bank flows, a redistribution ruled by some fiscal device) is *direct*. If both the ω are *ensembles* or classes in the partition of $\mathbb{X}$, then interaction is *indirect* or *mean-field*, which is the prototypical case of statistical mechanics.

When referred to income or wealth, such interactions have a peculiar characteristic. Withdrawing the $q\%$ from ω_P to transfer it to ω_R determines in the former an impoverishment effect which is more intensive than the enrichment effect on the latter. On the contrary, withdrawing $q\%$ from ω_R to transfer it to ω_P determines an enrichment effect on the latter which is more intensive than the impoverishment effect suffered by the former: if a poor inherits the heritage of the so-called *American uncle* he becomes rich; if a rich inherits the same, he may even not notice it and the more he is rich, the lower the effect if the heritage is not in the order of his wealth. Therefore, the lower the income level, the more effective is the effect of the

[6]This issue is directly tied to the *Pigou-Dalton transfer principle* (Pigou 1912; Dalton 1920) and subsequent refinements (Kolm 1969) as well as the notion of *mean-preserving spread* (Rothschild and Stiglitz 1970).

transfer. The second characteristic trait is therefore due to *interaction*, which has a very peculiar phenomenology and *a-symmetric* effects depending on the dimension of the entities in interaction.

A.6.4 Formal A-Simmetry: Concentration and Inequality

Then, Ω_n is a complex system of heterogeneous and interacting agents (HIA) characterized by two fundamental traits, both dominated by a form of a-symmetry: (i) heterogeneity implies structural a-symmetry, making the field of realizations differently populated by levels; (ii) interaction is dominated by *behavioral* a-symmetry. Heterogeneity and interaction are different sides of the same coin: one implies and reinforces the other (see Landini and Gallegati 2014).

These principles of a-symmetry combined together provide the key to understand the phenomenology of transferable quantities endowed with an ERUM as *strange quantities*, i.e. quantities obeying different probabilistic principles whose overlapping effects can be observed in the density of X, that is in the relation between the realization field $\mathbb{X} = [x_{\min}; x_{\max}]$ and the probability field $\mathbb{I} = [0; 1]$ of the implied distribution.

Complexity theory studies recognize that a further distinctive trait of HIA systems is that some observables with limited supports, which are transferable and endowed with an ERUM also have densities characterized by different probabilistic principles in different parts of their supports, often with right long tails (e.g. the dimension of cities, the size of firms, the degree of networks). Income and wealth are among such cases, and their strange behavior is widely recognized a regularity in different economies.

In these specific cases it seems that a threshold ξ exists in the realization field $\mathbb{X}$ such that $\forall X(\omega) \leq \xi$ the reference probabilistic principle is the *family of exponential distributions* $\mathcal{E}(\boldsymbol{\theta})$, characterized by a given granularity, whilst $\forall X(\omega) > \xi$ the reference probabilistic principle is the *family of power-law distributions* $\mathcal{P}(\boldsymbol{\theta})$, characterized by a different granularity.

Given this phenomenology, the probability field $\mathbb{I}$ is not *isotropic* w.r.t. partitions of the realization field $\mathbb{X}$. For instance, over the poor P and medium M income classes it may happen that even two closely adjacent income levels may be associated to very different densities of probability, whilst two other income levels which are even very far from each other within the rich income class R may be associated to almost the same density of probability. In such cases it is not easy, although possible, specifying a numeric transformation (or a partition, or both) of the realization field in such a way that the probability field becomes isotropic or, at least, ruled by a unique principle of proportionality as in the case of the uniform distribution.

In analogy with physics, it is like units $\omega : X(\omega) \leq \xi$ not obeying an *exclusion principle* so as to follow, like *bosons*, the Bose-Einstein statistics: i.e. there is no limit to the number of poor or medium incomes other than that imposed by the total number of units and total income. But it is also like units $\omega : X(\omega) > \xi$ obeying an *exclusion principle* such that, like *fermions*, they follow the Fermi-Dirac statistics: that is, there is a limit to the number of rich incomes due to the number of units and total income. This *a*-symmetry in obeying or not an exclusion principle determines the asymmetry of the distribution and represents a form of the previously discussed *relativism*.[7]

Putting together these two perspectives, if total income does not change sensibly—as in the static or short run cases—one may sample as many poor-medium incomes as they are if rich incomes do not change their condition becoming medium or poor. Ideally, if a rich donates all its income or wealth to poor-medium income people, even proportionally, he will become poor but some poor or medium agents will transit to higher income classes due to the effect of the received transfer. On the one hand, it is usually poor-medium income consumers that receive a wage by rich ones, to whom they pay back a share of the transfer in terms of purchasing. On the other hand, it is taxes paid by rich people that, *per capita*, contribute more to income redistribution to society in terms of services and other indirect transfers. It is such a continuous interaction of heterogeneous agents which maintains or increases concentration and inequality of income distribution. Therefore, interaction stimulates heterogeneity as heterogeneity determines interaction, and both participate to concentration and inequality which feeds the whole mechanism of transfers.

According to this perspective, the limit of total income is less relevant for poor agents than for rich ones: there can be $N - 1$ absolutely poor agents and one rich owning the whole amount (e.g. the Pharaoh), but if there is one agent in absolute poverty and the other $N - 1$ ones equally share the whole amount then terms like *rich* would be meaningless. The former is the extreme case of inequality, usually (and inappropriately) measured by means of the Gini concentration index: the Gini tells *how much* income is concentrated, not how it is concentrated across income classes, which is what matters for inequality and appropriately measured by the asymmetry of the Lorenz curve tied to the income distribution (see Gallegati et al. 2015). As a matter of fact, a given Gini may correspond to two different Lorenz curves with opposite asymmetry, whilst a given asymmetry measurement corresponds to one and only one Lorenz curve, even common to two income distributions.

However, under the implicit assumption that the income distribution does not change its symmetry profile, which is generally an empirical regularity, one may then indulge in the luxury of considering the Gini concentration as a standard measure of inequality. Therefore, as a consequence, one may conclude that the higher the concentration the higher the inequality because it is empirically observed across economies that very few rich people concentrate the hugest share of total income: i.e. there is always a restricted share of rich agents obeying the exclusion principle while the wider one of poor-medium agents does not.

---

[7]Of course, this is a weak and impressionistic analogy, only having the aim at tracing the intuition behind. More rigorously, one should consider a wide-sense or generalized exclusion principle, for which *bosons* and *fermions* are particular cases. In this more general perspective one should consider the fractional statistics of *anyons*, particles which may have a *boson*-like or *fermion*-like behavior or they may even perform a mixed behavior depending on the conditions. Giorgio Kaniadakis is gratefully acknowledged for having suggested readings in this field.

This implication has also a second interpretation: the higher the inequality, the wider the set of realizations ruled by $\mathcal{E}(\boldsymbol{\theta})$, hence the more restricted the set of realizations ruled by $\mathcal{P}(\boldsymbol{\theta})$. Stated otherwise, the few incomes ruled by $\mathcal{P}(\boldsymbol{\theta})$ accumulate the hugest share of total income, while the many ones ruled by $\mathcal{E}(\boldsymbol{\theta})$ accumulate the smallest one: clearly, the former are the rich incomes, whereas the latter the poor-medium ones.

A.7 Deform to Go Strange: Mathematics of the κ-Generalized Model

The problem is as follows: it seems that for a strange quantity X (e.g. income or wealth) there exists some threshold value ξ for which $X \sim \mathcal{E}(\boldsymbol{\theta})$, if $X(\omega) \leq \xi$, whereas for $X(\omega) > \xi$ the quantity follows $X \sim \mathcal{P}(\boldsymbol{\theta})$. Is it possible to reconcile such different probabilistic characters into a single model for $X(\omega) \in \mathbb{X}$ unifying the principles underlying the strange behavior of X? The answer, of course, is positive.

As mentioned in this book, there are different models reconciling with some of the strange properties of X, most of which belong to the family of the generalized beta distribution—embedded, in turn, in the exponential family: the generalized beta of the second kind (GB2), Singh-Maddala and Dagum type I are the most famous while other distributions, such as the Weibull, are less versatile to describe income and wealth—although they still satisfy the weak Pareto law. Besides all such distributions for income and wealth, the κ-generalized model—which is different from the others—performs pretty well: it does not belong to a family, rather it defines a new family of distributions with dependence both on parameters and, more importantly, on the choice of the specifying function and generator of the deformation. Indeed, while all the previous distributions operate on the support $\mathbb{X}$ of X, eventually manipulated by means of some *numerical transformations*, the κ-generalized model operates on a *deformation* of $\mathbb{X}$ and functions operating on it so as to grasp the strange behavior of strange quantities—not necessarily income and wealth.

The following sections provide a first introduction to the mathematics of deformations needed to specify the κ-generalized distribution. Among the above-mentioned distributions, the κ-generalized model is the only one which formally and explicitly embeds both the exponential and the power-law shapes in a single functional form: therefore, when operating on that portion of $\mathbb{X}$ ruled by probabilistic principles of the exponential family $\mathcal{E}(\boldsymbol{\theta})$ (no exclusion principle) the κ-generalized model replicates fairly well the underlying data, and the same good performance holds when it operates on that portion of the income or wealth support ruled by probabilistic principles of the power-law family $\mathcal{P}(\boldsymbol{\theta})$ (exclusion principle), thus capturing both the stochastic and relativistic aspects mentioned so far.

A.7.1 Notes on the Generalized κ-Deformed Mathematics

A.7.1.1 Self-Duality

Every function $f : \mathbb{R} \rightarrow \mathbb{R}$ is decomposable in the sum of its *e*ven and *o*dd parts as follows

$$\begin{cases} f(x) = f_e(x) + f_o(x), \\ f(-x) = f_e(-x) + f_o(-x) = f_e(x) - f_o(x). \end{cases} \tag{A.56}$$

By adding and subtracting previous expressions, one gets

$$\begin{cases} \text{addition} : \Sigma(x) = f(x) + f(-x) = 2f_e(x), \\ \text{subtraction} : \Delta(x) = f(x) - f(-x) = 2f_o(x). \end{cases} \tag{A.57}$$

Therefore, the even and odd parts of $f(x)$ are defined by

$$f_e(x) = \frac{1}{2}\Sigma(x) \quad \text{and} \quad f_o(x) = \frac{1}{2}\Delta(x). \tag{A.58}$$

A function $f(x)$ is *self-dual* if

$$f(x)\, f(-x) = 1. \tag{A.59}$$

Self-duality implies

$$\begin{cases} f(x)\, f(-x) = [f_e(x) + f_o(-x)][f_e(x) - f_o(x)] \\ \qquad\qquad\quad = f_e^2(x) - f_o^2(x) = 1, \\ f(x)\, f(-x) = 1 \Rightarrow f_e(x) = \sqrt{1 + f_o^2(x)}. \end{cases} \tag{A.60}$$

Therefore, substituting (A.60) into (A.56) gives

$$f(x) = \sqrt{1 + f_o^2(x)} + f_o(x), \tag{A.61}$$

which is the *odd representation* of every self-dual function $f(x)$. Furthermore, since $\text{arcsinh}(u) \equiv \ln\left(\sqrt{1+u^2} + u\right)$, then

$$\ln[f(x)] = \text{arcsinh}[f_o(x)] \Leftrightarrow f(x) = \exp\{\text{arcsinh}[f_o(x)]\} \tag{A.62}$$

is the *generative representation*, where $f_o(x)$ is the *generator*—an odd function.

Deformations involved in the following are based on (A.61) and (A.62). Moreover, it is worth noticing that $s(x) = \exp(x)$ is self-dual and, according to (A.57), $s_e(x) = \left(e^x + e^{-x}\right)/2 = \cosh(x)$ and $s_o(x) = \left(e^x - e^{-x}\right)/2 = \sinh(x)$.

A.7.1.2 The Generalized κ-Deformation

Consider the random variable $X : \Omega_n \to \mathbb{X} = [x_{\min}; x_{\max}]$, $0 < x_{\min} \ll x_{\max} < +\infty$, representing income or wealth. Let $G \in \mathcal{C}^{\infty}(\mathbb{R})$ be a strictly increasing invertible odd function such that $g(x) \approx x$ as $x \to x_{\min}$. Let also $\kappa \in \mathbb{R}$ and define

$$g_{\kappa}(x) = G(\kappa x) . \tag{A.63}$$

Being κ the *deformation parameter* of the generative representation, the function

$$X_{\kappa} \equiv \frac{1}{\kappa} \operatorname{arcsinh}\left[g_{\kappa}(X)\right] \tag{A.64}$$

is the *κ-deformation of* X induced by the generator G if:

- it is a 1 to 1 order-preserving mapping which deforms $\mathbb{X}$ into $\mathbb{X}_{\kappa}$;
- $X_{\kappa} \approx X$ both for $x \to x_{\min}$ and for $\kappa \to 0$, being $\min \mathbb{X}_{\kappa} = x_{\min}$ and $X_0 = X$;
- $X_{-\kappa} = X_{\kappa}$ because $\operatorname{arcsinh}(u)$ is odd, therefore κ can be restricted on $\mathbb{R}_+$;
- the function

$$X^{\kappa} \equiv \frac{1}{\kappa} G^{-1}\left[\sinh(\kappa X)\right] \tag{A.65}$$

  is such that $(X^{\kappa})_{\kappa} = (X_{\kappa})^{\kappa} = X$ and defines the inverse deformation of X_{κ}.

In the present context it has been discussed that $\mathbb{X}$ is not everywhere dominated by the same statistical principle, rather there are different ones characterizing different portions of the realization field of X, which is the reason why it is a strange quantity. The κ-deformation aims at reconciling these principles into a unified one in such a way that the strange behavior of X on $\mathbb{X}$ is deformed into the more regular behavior of X_{κ} on $\mathbb{X}_{\kappa}$ without modifying the ordering. In terms of the discussed weak analogy w.r.t. the Galilean relativity (classic), where space and time are absolute and distinct, and Einstein relativity (relativistic), where space and time melt into the space-time, the deformation plays a role similar to the the Lorentz transform: as for $c \to +\infty$ in the Lorentz transform the Einstein relativity meets the Galilean one (i.e. at subluminal velocities), if $\kappa \to 0$ in the κ-deformation then the deformed quantity meets the strange one. Hence, the more the quantity X behaves regularly on its support, which means it is everywhere ruled by the same statistical principle, the more the deformation parameter approaches 0. On the other hand, if X behaves strangely, i.e. with coexisting different statistical principles on its support, the more the deformation parameter is different from zero.

Figure A.1 shows the effect of the deformation on $s(X) = (X/\beta)^{\alpha}$. Original data have been "strangely" simulated.[8] Panel (a) plots the X simulated values and their $s(X)_{\kappa}$ deformation with two different values for the deformation parameter: $\kappa = 0.1$ and $\kappa = 10$. The deformation (A.64) involves a *specializing* function $s(x) = (x/\beta)^{\alpha}$

[8] More specifically, $n = 1,000,000$ random numbers X were obtained by multiplying by 1, 000 the cubic root of ratio between $A \sim \text{LN}[\ln(40,000), 0.9]$ and $B \sim \text{LN}[\ln(20,000), 1.8]$.

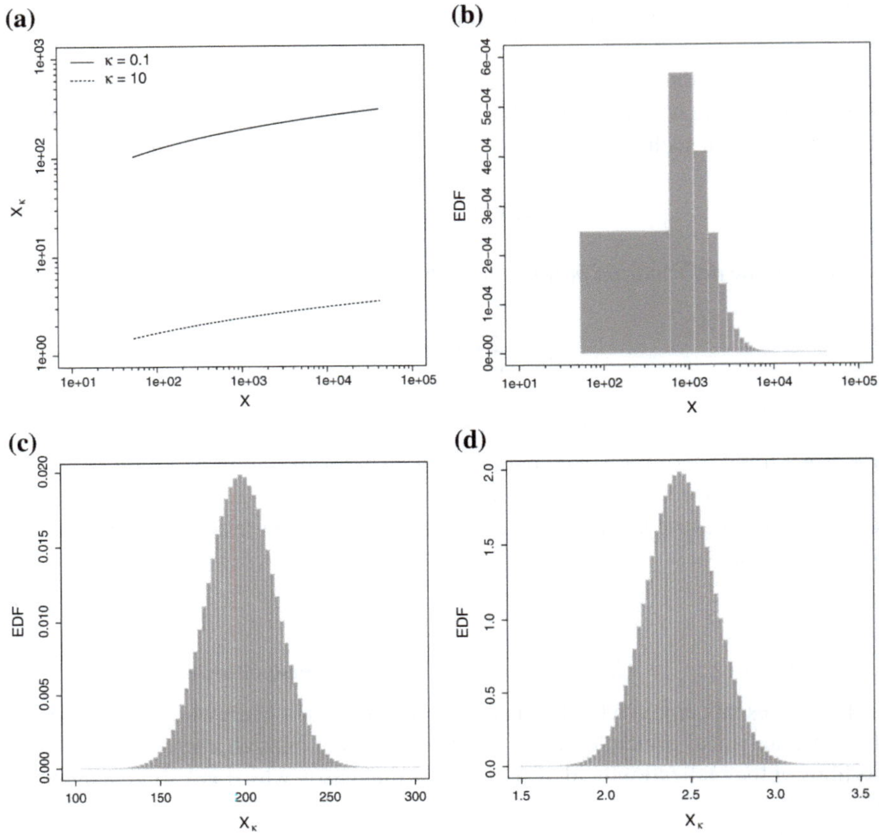

Fig. A.1 Effect of the generalized κ-deformation applied to the specializing transformation $s(x) = (X/\beta)^\alpha$, $\alpha = 3$, $\beta = 1$ and $\kappa = 0.1, 10$: **a** strange versus deformed; **b** EDF of the strange X; **c** EDF of the deformed X_κ, $\kappa = 0.1$; **d** EDF of the deformed X_κ, $\kappa = 10$

with $\alpha = 3$ and $\beta = 1$, while the odd *generator* is $G[\kappa s(x)] = \kappa (x/\beta)^\alpha$. As it can be seen, the empirical density function (EDF) of the strange X exhibits a long tail to the right and a huge bunch of observations around the expected value. On the other hand, after the deformation is applied, the EDF becomes more regular: the higher the value of κ, the more the deformation restricts the width of the realization field while reshaping the EDF into an apparently regular bell-shaped curve. Notice that this is not the κ-generalized distribution, but only the EDF of the strange quantity X deformed according to (A.64) and applied to $s(X) = (X/\beta)^\alpha$.

Of course, there is not only one deformation to deal with; different invertible specializing functions $s(x)$ can be involved as initial transformations, such as for instance rescaling; moreover, one can also consider different generators G. Therefore, (A.64) is a *generalized* deformation.

A.7.1.3 Generalized κ-Deformed Operators

To specify the κ-generalized model we have to introduce the mathematics of deformations. By assuming the simplest specializing and generator functions discussed in the following, Kaniadakis (2001, 2002, 2005) developed a complete deformed mathematics and traced the way to extend results toward more general derivations. Therefore, the following does not develop a complete mathematics of the generalized κ-deformation but it only concerns those aspects related to the derivation of the κ-generalized model.

According to definition (A.64) the deformed quantity is a function of the strange original one. Therefore, the *generalized κ-deformed differential* is given by

$$\mathrm{d}x_\kappa \equiv \frac{1}{\kappa}\frac{G'(\kappa x)}{\sqrt{1+G^2(\kappa x)}}\,\mathrm{d}x. \tag{A.66}$$

The deformed differential gives the *generalized κ-deformed derivative*

$$\frac{\mathrm{d}}{\mathrm{d}x_\kappa} \equiv \kappa\frac{\sqrt{1+G^2(\kappa x)}}{G'(\kappa x)}\frac{\mathrm{d}}{\mathrm{d}x}. \tag{A.67}$$

To specify the κ-deformed exponential one can resort to the odd representation (A.61). By considering the operator $g_\kappa(x)$ defined in (A.63), a function originating from this generator exists, say $E_\kappa(x) = [e_\kappa(x)]^{1/\kappa} > 0$, such that

$$E_\kappa(x) = \sqrt{1+g_\kappa^2(x)} + g_\kappa(x) \Leftrightarrow e_\kappa(x) = \left[\sqrt{1+g_\kappa^2(x)} + g_\kappa(x)\right]^{\frac{1}{\kappa}}. \tag{A.68}$$

By substituting (A.63), it then follows that

$$e_\kappa(x) = \left[\sqrt{1+G^2(\kappa x)} + G(\kappa x)\right]^{\frac{1}{\kappa}}. \tag{A.69}$$

Since $g_\kappa(x) = G(\kappa x)$, by applying logarithm on both sides it gives

$$\ln e_\kappa(x) = \frac{1}{\kappa}\ln\left[\sqrt{1+G^2(\kappa x)} + G(\kappa x)\right] = \frac{1}{\kappa}\,\mathrm{arcsinh}\,[g_\kappa(x)], \tag{A.70}$$

hence

$$e_\kappa(x) = \exp\left\{\frac{1}{\kappa}\,\mathrm{arcsinh}\,[g_\kappa(x)]\right\} = \exp(x_\kappa) \tag{A.71}$$

fulfills the following properties that characterize the exponential function:

- self-duality, that is

$$e_\kappa(x)\,e_\kappa(-x) = 1; \tag{A.72}$$

- it coincides with its κ-deformed derivative, i.e.

$$\frac{\mathrm{d}}{\mathrm{d}\,x_\kappa} e_\kappa\,(x) = e_\kappa\,(x)\,. \tag{A.73}$$

Therefore, by setting $e_\kappa \equiv \exp_\kappa$ yields

$$\begin{aligned}\exp_\kappa\,(x) &\equiv \left[\sqrt{1+G^2\,(\kappa x)} + G\,(\kappa x)\right]^{\frac{1}{\kappa}} \\ &= \exp\left\{\frac{1}{\kappa}\,\mathrm{arcsinh}\,[g_\kappa\,(x)]\right\} \\ &= \exp\,(x_\kappa)\,,\end{aligned} \tag{A.74}$$

deformed exponential of fulfills previous properties and defines the κ*-deformed exponential function*. Figure A.2 shows the κ-deformed exponential of $s\,(x) = -\,(x/\beta)^\alpha$ with $g_\kappa\,(x) = \kappa s\,(x)$ for X simulated as in Fig. A.1: as it can be seen, the specialized and deformed exponential embeds both the exponential and the power-law characters.

Inverse to the deformed exponential is the κ*-deformed logarithm*, which can be defined by means of the generalized inverse κ-deformation. Indeed, by substituting for $y = \ln\,(x)$ in (A.65), then $y^\kappa = [\ln\,(x)]^\kappa$ gives

$$[\ln\,(x)]^\kappa = \frac{1}{\kappa} G^{-1}\,\{\sinh\,[\kappa \ln\,(x)]\}\,. \tag{A.75}$$

Since $\sinh\,(u) = \left[e^u - e^{-u}\right]/2$, let $u = \kappa \ln\,(x)$; then simple algebra gives

$$\sinh\,[\kappa \ln\,(x)] = \frac{x^\kappa - x^{-\kappa}}{2} \Leftrightarrow \kappa \ln\,(x) = \mathrm{arcsinh}\left(\frac{x^\kappa - x^{-\kappa}}{2}\right), \tag{A.76}$$

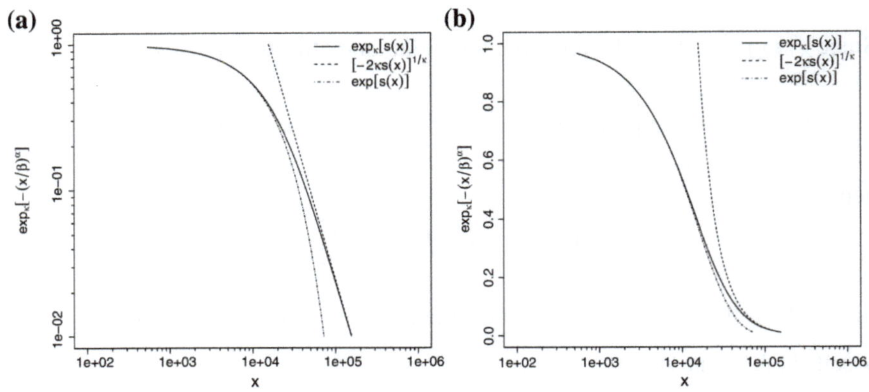

Fig. A.2 The generalized κ-deformed exponential of $s\,(x) = -\,(x/\beta)^\alpha$ with $g_\kappa\,(x) = \kappa s\,(x)$, $\alpha = 1$, $\beta = \bar{X}$ and $\kappa = 1$; **a** log-log scale; **b** linear-log scale

hence

$$[\ln(x)]^{\kappa} = \frac{1}{\kappa} G^{-1}\left(\frac{x^{\kappa} - x^{-\kappa}}{2}\right). \tag{A.77}$$

On the other hand, if the deformed logarithm is to be the inverse of the deformed exponential, then $\ln_{\kappa}\left[\exp_{\kappa}(x)\right] = \exp_{\kappa}\left[\ln_{\kappa}(x)\right] = x$. Define

$$y = \ln_{\kappa}(x) \Leftrightarrow \exp_{\kappa}(y) = x \tag{A.78}$$

such that

$$\exp_{\kappa}(y) = \exp\left\{\frac{1}{\kappa}\,\mathrm{arcsinh}\left[G(\kappa y)\right]\right\}, \tag{A.79}$$

which means

$$\sinh\left\{\kappa \ln\left[\exp_{\kappa}(y)\right]\right\} = G(\kappa y). \tag{A.80}$$

By setting $u = \kappa \ln\left[\exp_{\kappa}(y)\right]$, then $\sinh(u) = \left[e^{u} - e^{-u}\right]/2$ gives

$$\frac{1}{\kappa} G^{-1}\left(\frac{x^{\kappa} - x^{-\kappa}}{2}\right) = y. \tag{A.81}$$

Therefore, due to (A.77) and (A.78), the κ-deformed logarithm is

$$\ln_{\kappa}(x) \equiv \frac{1}{\kappa} G^{-1}\left(\frac{x^{\kappa} - x^{-\kappa}}{2}\right) = [\ln(x)]^{\kappa}. \tag{A.82}$$

A.7.1.4 The Kaniadakis Deformation

The Kaniadakis deformation is defined by the following generator

$$G(\kappa x) \equiv \kappa x, \tag{A.83}$$

so that the deformation (A.64) reads as

$$X_{\kappa} = \frac{1}{\kappa}\,\mathrm{arcsinh}(\kappa X), \tag{A.84}$$

which reduces to X as $\kappa \to 0$, while the inverse is

$$X^{\kappa} = \frac{1}{\kappa}\sinh(\kappa X). \tag{A.85}$$

Accordingly, the differential (A.66) becomes

$$\mathrm{d}\, x_{\kappa} = \frac{\mathrm{d}\, x}{\sqrt{1+\kappa^2 x^2}}. \tag{A.86}$$

Notice that, as far as the strange observable increases, the deformed differential decreases. That is, a small change of the strange quantity X has not the same effect everywhere on the realization field; the deformed differential grasps this effect and becomes less and less effective as far as the values increase, such as in the case of top incomes. Moreover, it reduces to the standard differential as $\kappa \to 0$, that is when no deformation is required.

Under the assumption (A.83), the derivative (A.67) reads as

$$\frac{\mathrm{d}}{\mathrm{d}\, x_{\kappa}} = \sqrt{1+\kappa^2 x^2}\frac{\mathrm{d}}{\mathrm{d}\, x} \tag{A.87}$$

and, of course, it coincides with the standard derivative as $\kappa \to 0$.

The Kaniadakis κ-deformed exponential and logarithm follow by substituting for (A.83) into (A.74) and (A.82), respectively, i.e.

$$\begin{cases} \exp_{\kappa}(x) = \left(\sqrt{1+\kappa^2 x^2} + \kappa x\right)^{\frac{1}{\kappa}}, \\ \ln_{\kappa}(x) = \dfrac{x^{\kappa} - x^{-\kappa}}{2\kappa}. \end{cases} \tag{A.88}$$

Few main properties are now worth mentioning:

- As far as X approaches top incomes and wealth holdings, that is for $x \to +\infty$, then $\forall \kappa > 0$ the Kaniadakis exponential and logarithm asymptotically converge as follows
$$\exp_{\kappa}(x) \to (2\kappa x)^{1/\kappa} \quad \text{and} \quad \ln_{\kappa}(x) \to \frac{1}{2\kappa} x^{\kappa}. \tag{A.89}$$
- As far as X approaches poorest incomes and wealth holdings, that is for $x \to 0^+$, then $\forall \kappa > 0$ the Kaniadakis exponential and logarithm asymptotically converge as
$$\exp_{\kappa}(x) \to \exp(x) \quad \text{and} \quad \ln_{\kappa}(x) \to -\frac{1}{2\kappa} x^{-\kappa}. \tag{A.90}$$
- Both deformed functions converge to the standard ones as $\kappa \to 0$, i.e.
$$\exp_{\kappa}(x) \to \exp(x) \quad \text{and} \quad \ln_{\kappa}(x) \to \ln(x)\,; \tag{A.91}$$
thus, if the quantity X is not strange no deformation is required.

A more detailed presentation of the Kaniadakis deformed mathematics is developed in Kaniadakis (2002, 2005).

A.7.2 The κ-Generalized Distribution

Assume for the sake of exposition that X is a strange quantity representing income with support $\mathbb{X} = [x_{\min}; x_{\max}]$, where $0 \ll x_{\min} \ll x_{\max} < +\infty$. Also, consider the simplest but generic transformation beyond the identity $s_0(x) = x$, say $s_1(x) = bx$, where $b \neq 0, 1$. As discussed in previous sections, the order of magnitude may be very different for realizations of X, hence a first adjustment may be a rescaling transformation such as

$$s_1(x) = \frac{x}{\beta} \quad \beta > 0, \tag{A.92}$$

which is invertible and odd: suitable candidates for β are $x_{\min}$, $\bar{X}$ or $x_{\max}$. As it is clear, (A.92) is nothing but a proportional rescaling function tuned by the scale parameter β. To improve the regularization of the strange quantity X, a further transformation just more complex than (A.92) can be introduced, that is

$$s_2(x) = \left(\frac{x}{\beta}\right)^{\alpha}, \quad \alpha, \beta > 0. \tag{A.93}$$

The specializing transformation (A.93) introduces both a rescaling and a power, which reads as a filtering-root if $\alpha \in (0, 1)$. Moreover, notice that

$$y = s_2(x) \Rightarrow \mathrm{d}\,y = \frac{\alpha}{\beta}\left(\frac{x}{\beta}\right)^{\alpha-1} \mathrm{d}\,x. \tag{A.94}$$

Therefore, if a small change in x implies an effect which is not everywhere the same, then $y = s_2(x)$ might be a starting point to grasp the strange character of X.

Given the specializing transformation y of the original x, consider the generator

$$g_\kappa(y) = G(\kappa y) = \kappa y, \quad \kappa \in [0, 1), \tag{A.95}$$

to construct the κ-deformation of y of the transformed x as

$$y_\kappa = \frac{1}{\kappa} \operatorname{arcsinh}(\kappa y) \Rightarrow \mathrm{d}\,y_\kappa = \frac{\mathrm{d}\,y}{\sqrt{1 + \kappa^2 y^2}}. \tag{A.96}$$

Notice that as far as the transformed y increases, the implied change decreases. By means of the same generator define then the following operators

$$\exp_\kappa(\pm y) = \sqrt{1 + \kappa^2 y^2} \pm \kappa y. \tag{A.97}$$

So far, it has been observed that the information content is lower for common events, while it is higher for more unusual ones, hence the order-preserving deformed exponential $y = s_2(x)$ can be suitably modified according to the principle $-y = -s_2(x)$ to induce the following *κ-deformed exponential distribution*

$$U_\kappa(y) = 1 - \exp_\kappa(-y) \rightarrow \begin{cases} 0 : y \rightarrow 0^+ \Leftrightarrow x \rightarrow 0^+, \\ 1 : y \rightarrow +\infty \Leftrightarrow x \rightarrow +\infty. \end{cases} \tag{A.98}$$

By using the derivative as defined in (A.87), and since the derivative of the exponential is the exponential itself also in the deformed case as in (A.73), the density function is

$$u_\kappa(y) = \frac{\mathrm{d}\,U_\kappa(y)}{\mathrm{d}\,y} = \frac{\exp_\kappa(-y)}{\sqrt{1+\kappa^2 y^2}}, \tag{A.99}$$

where $\exp_\kappa(-y) : y = s_2(x)$ is shown in Fig. A.2: as discussed, this function grasps either the exponential and the power-law profiles, hence it seems to be a good candidate to develop a distribution for a strange quantity. On the other hand, beyond being deformed, it is still an exponential function: if it were $\kappa = 0$, then $\exp_0(-y) = \exp\left[-(x/\beta)^\alpha\right]$, hence $f(x; \alpha, \beta, \kappa = 0) = \beta^{-\alpha} \exp\left[-(x/\beta)^\alpha\right]$ would be an exponential distribution.

Notice that by substituting for (A.94) in (A.99) gives

$$u_\kappa\left[(x/\beta)^\alpha\right] = \frac{\mathrm{d}\,U_\kappa\left[(x/\beta)^\alpha\right]}{\frac{\alpha}{\beta}\left(\frac{x}{\beta}\right)^{\alpha-1}\mathrm{d}\,x} = \frac{\exp_\kappa\left[-(x/\beta)^\alpha\right]}{\sqrt{1+\kappa^2(x/\beta)^{2\alpha}}}. \tag{A.100}$$

The function

$$\begin{aligned} f(x; \alpha, \beta, \kappa) &= \frac{\mathrm{d}\,U_\kappa\left[(x/\beta)^\alpha\right]}{\mathrm{d}\,x} \\ &= \frac{\alpha}{\beta}\left(\frac{x}{\beta}\right)^{\alpha-1} u_\kappa\left[(x/\beta)^\alpha\right] \\ &= \frac{\alpha}{\beta}\left(\frac{x}{\beta}\right)^{\alpha-1} \frac{\exp_\kappa\left[-(x/\beta)^\alpha\right]}{\sqrt{1+\kappa^2(x/\beta)^{2\alpha}}} \end{aligned} \tag{A.101}$$

defines the *κ-generalized density function* as the regular derivative of the κ-deformed exponential distribution of the transformed strange observable X. Therefore, a quantity X obeying (A.101) is said to be κ-generalized distributed with shape parameter $\alpha > 0$, scale parameter $\beta > 0$ and deformation parameter $\kappa \in [0, 1)$.

Few remarks about the asymptotic properties of the κ-generalized model are worth stressing:

- As $\kappa \rightarrow 0^+$, the κ-generalized model behaves *classically*: indeed, κ-gen $(\alpha, \beta, \kappa = 0) = \text{Wei}(\alpha, \beta)$ and $\text{Wei}(\alpha = 1, \beta) = \text{Exp}(\beta)$, then the standard exponential distribution is met by the κ-generalized with shape parameter $\alpha = 1$ and deformation parameter $\kappa = 0$. Hence, if the observable X is κ-generalized distributed, but not strange, then the true distribution belongs to the exponential family.

- As $0 < \kappa < 1$, the κ-generalized model behaves *relativistically*:
  - For $x \to 0^+$, meaning $x \to x_{\text{min}}^+$, $X \sim \mathcal{E}(\boldsymbol{\theta})$: the κ-generalized model at the lower tail of the income or wealth distribution follows a standard distribution of the exponential family.
  - For $x \to +\infty$, meaning $x \to x_{\text{max}}^-$, $X \sim \mathcal{P}(\boldsymbol{\theta})$: the κ-generalized model converges to a Pareto distribution $f(x; \alpha, \beta, \kappa) = (\alpha/\kappa)\left[\beta(2\kappa)^{-1/\alpha}\right]^{\alpha/\kappa} x^{-(\alpha/\kappa+1)}$.

A.8 Concluding Remarks

As a conclusion to this appendix, a brief answer to the question suggested by the title is proposed to summarize what developed in previous sections and through the whole book.

While noticing that distributions of the power-law family are almost ubiquitous in several fields of science, it has also been recognized that their appearance is often contextual with other laws, mainly those belonging to the family of exponential distributions. This coexistence has been termed here *strangeness of observables*. Since it is observed in very different fields, e.g. physics and economics, it seems there is no unique principle explaining this coexistence; rather, there can be different principles common to different fields that can be explained with a unified interpretation.

To describe such an emergent property, the perspective of complex systems made of many heterogeneous and interacting agents is assumed, and the suitable methodology involved is *statistical mechanics*. Moreover, to explain the strange character of stochastic quantities behaving relativistically, that is according to exponential laws below some threshold and power laws above it, the *mathematics of deformations* is involved too.

According to this twofold methodology, an explanation for strangeness in complex systems is that the realization field (i.e. the support) of a *strange quantity* is not isotropic with respect to the associated probability field. This phenomenon is found at the roots of the a-symmetry property of strange quantities: structural a-symmetry due to heterogeneity and granularity, behavioral a-symmetry due to interaction and transferability, and formal a-symmetry due to concentration and inequality are isolated as those principles which may be commonly shared in different fields.

The main cause of a-symmetry of a *strange quantity* is that there are portions of its support at low orders of magnitude in which observation units do not obey an *exclusion principle*, as if they were *bosons*, and the exponential family of distributions well describes them, while at higher orders of magnitude units obey an *exclusion principle*, as if they were *fermions*, and the power-law family of distributions well describes these others.

Melting statistical mechanics and the mathematics of deformations, it is shown that the regular derivative of the κ-deformed exponential distribution of a transformed strange observable defines the κ-generalized density function. This parametric model does not belong to any class of distributions involved in the analysis of income and wealth; rather, it defines a new class of distributions which explicitly embeds in a unified density function both the exponential and the power-law behaviors. Moreover, if estimates of the deformation parameter κ are about to be zero, whether if not zero, then the model behaves *classically* because the underlying quantity is not strange, and models like the Weibull distribution can be found as particular cases. On the other hand, if estimates are such that $0 < \kappa < 1$, then the model behaves *relativistically* because the quantity is strange. In such cases, at low orders of magnitude on the support the exponential family of distributions emerges whilst at high orders of magnitude the power-law family provides a good fit.

Therefore, not depending on the field of application, that is even beyond the horizon of economics and physics, the κ-generalized model is suitable to describe *strange* and *regular* quantities with a high degree of fitness while receiving a theoretical motivation which grows upon the phenomenology of the underlying observable.

Appendix B
Supplementary Tables

Table B.1 Maximum likelihood estimates for the Singh-Maddala, Dagum type I and κ-generalized models of income distribution: all LIS countries in the most recent years

Code[a]	Model[b]	Parameters[c]		
		a (α)	b (β)	q, p, κ
AU10	SM	2.401	56,582	1.723
		(0.020)	(937)	(0.046)
	D	3.398	52,303	0.623
		(0.037)	(509)	(0.012)
	κ-gen	2.173	49,633	0.652
		(0.019)	(256)	(0.017)
AT04	SM	3.438	21,577	1.193
		(0.069)	(497)	(0.061)
	D	3.999	22,051	0.782
		(0.068)	(296)	(0.026)
	κ-gen	3.093	22,248	0.901
		(0.057)	(174)	(0.038)
BE00	SM	3.339	718,001	1.224
		(0.081)	(19,883)	(0.072)
	D	3.922	728,417	0.778
		(0.144)	(22,890)	(0.060)
	κ-gen	2.937	742,651	0.819
		(0.077)	(8,821)	(0.051)

(continued)

F. Clementi and M. Gallegati, *The Distribution of Income and Wealth*,
New Economic Windows, DOI 10.1007/978-3-319-27410-2

Table B.1 (continued)

Code[a]	Model[b]	Parameters[c]		
		a (α)	b (β)	q, p, κ
BR13	SM	2.041	14,474	1.110
		(0.009)	(127)	(0.013)
	D	2.192	14,390	0.906
		(0.010)	(109)	(0.010)
	κ-gen	1.826	15,938	0.949
		(0.008)	(41)	(0.009)
CA10	SM	2.545	47,514	1.691
		(0.020)	(679)	(0.041)
	D	3.640	45,109	0.606
		(0.040)	(414)	(0.013)
	κ-gen	2.306	42,372	0.660
		(0.017)	(174)	(0.014)
CN02	SM	1.331	11,124	2.272
		(0.010)	(367)	(0.070)
	D	1.823	6,185	0.758
		(0.021)	(137)	(0.019)
	κ-gen	1.177	7,118	0.448
		(0.012)	(73)	(0.022)
CO10	SM	1.703	6,025,950	1.356
		(0.022)	(191,567)	(0.048)
	D	2.086	5,765,774	0.757
		(0.022)	(97,812)	(0.016)
	κ-gen	1.558	5,801,406	0.845
		(0.015)	(46,026)	(0.020)
CZ10	SM	3.824	223,810	1.042
		(0.062)	(3,606)	(0.041)
	D	3.964	225,574	0.940
		(0.063)	(3,305)	(0.035)
	κ-gen	3.385	240,810	0.985
		(0.051)	(1,444)	(0.033)
DK10	SM	3.166	295,007	1.902
		(0.014)	(1,921)	(0.027)
	D	5.086	283,733	0.506
		(0.033)	(1,012)	(0.006)
	κ-gen	2.895	257,397	0.605
		(0.011)	(418)	(0.007)

(continued)

Table B.1 (continued)

Code[a]	Model[b]	Parameters[c]		
		a (α)	b (β)	q, p, κ
EG12	SM	1.964	7,905	1.122
		(0.017)	(131)	(0.022)
	D	2.165	8,116	0.854
		(0.020)	(92)	(0.012)
	κ-gen	1.774	8,611	0.963
		(0.020)	(72)	(0.023)
EE10	SM	2.367	142,892	1.893
		(0.033)	(4,266)	(0.092)
	D	3.383	123,392	0.639
		(0.067)	(2,122)	(0.023)
	κ-gen	2.158	119,173	0.617
		(0.037)	(1,206)	(0.035)
FI10	SM	3.473	26,293	1.283
		(0.048)	(458)	(0.050)
	D	4.080	25,822	0.803
		(0.065)	(342)	(0.027)
	κ-gen	3.050	26,746	0.782
		(0.039)	(147)	(0.026)
FR10	SM	3.017	24,211	1.379
		(0.031)	(349)	(0.040)
	D	3.923	24,607	0.668
		(0.043)	(209)	(0.013)
	κ-gen	2.737	23,682	0.801
		(0.027)	(109)	(0.019)
DE10	SM	3.170	22,156	1.251
		(0.036)	(317)	(0.037)
	D	3.709	22,028	0.798
		(0.047)	(250)	(0.021)
	κ-gen	2.816	22,662	0.830
		(0.032)	(120)	(0.024)
GE13	SM	2.306	3,705	1.125
		(0.061)	(170)	(0.075)
	D	2.357	3,323	1.052
		(0.048)	(95)	(0.044)
	κ-gen	2.018	4,050	0.897
		(0.055)	(73)	(0.060)

(continued)

Table B.1 (continued)

Code[a]	Model[b]	Parameters[c]		
		a (α)	b (β)	q, p, κ
GR10	SM	2.311	18,353	1.889
		(0.029)	(415)	(0.068)
	D	3.613	17,103	0.533
		(0.084)	(300)	(0.021)
	κ-gen	2.124	15,224	0.625
		(0.030)	(127)	(0.025)
GT06	SM	1.651	17,339	1.387
		(0.020)	(578)	(0.050)
	D	1.929	14,754	0.860
		(0.022)	(298)	(0.020)
	κ-gen	1.463	16,875	0.753
		(0.016)	(164)	(0.023)
HU12	SM	2.773	1,905,113	1.769
		(0.059)	(60,641)	(0.106)
	D	3.993	1,769,296	0.612
		(0.152)	(51,737)	(0.044)
	κ-gen	2.527	1,679,619	0.648
		(0.068)	(21,854)	(0.051)
IS10	SM	3.717	3,896,441	1.298
		(0.068)	(77,056)	(0.061)
	D	4.640	3,983,394	0.703
		(0.122)	(69,319)	(0.035)
	κ-gen	3.316	3,921,461	0.809
		(0.076)	(35,421)	(0.046)
IN04	SM	1.476	23,278	1.818
		(0.009)	(449)	(0.034)
	D	2.084	18,861	0.644
		(0.015)	(197)	(0.009)
	κ-gen	1.379	17,482	0.708
		(0.010)	(108)	(0.014)
IE10	SM	2.679	29,804	1.785
		(0.042)	(825)	(0.090)
	D	3.764	26,895	0.644
		(0.069)	(429)	(0.022)
	κ-gen	2.433	26,052	0.643
		(0.053)	(304)	(0.046)

(continued)

Table B.1 (continued)

Code[a]	Model[b]	Parameters[c]		
		a (α)	b (β)	q, p, κ
IL12	SM	1.898	143,368	2.511
		(0.021)	(4,943)	(0.115)
	D	3.110	107,847	0.537
		(0.053)	(1,648)	(0.016)
	κ-gen	1.745	96,929	0.481
		(0.021)	(810)	(0.023)
IT10	SM	2.316	21,874	1.964
		(0.026)	(494)	(0.071)
	D	3.648	19,877	0.535
		(0.055)	(241)	(0.014)
	κ-gen	2.145	17,749	0.622
		(0.028)	(128)	(0.023)
JP08	SM	2.742	3,619,224	1.570
		(0.052)	(104,584)	(0.082)
	D	3.796	3,530,485	0.628
		(0.102)	(76,901)	(0.032)
	κ-gen	2.498	3,344,901	0.720
		(0.052)	(34,759)	(0.040)
LU10	SM	3.289	41,175	1.309
		(0.049)	(742)	(0.049)
	D	3.872	39,905	0.808
		(0.080)	(783)	(0.038)
	κ-gen	2.905	41,426	0.794
		(0.051)	(332)	(0.037)
MX12	SM	1.729	66,477	1.530
		(0.024)	(2,258)	(0.060)
	D	2.314	62,812	0.665
		(0.038)	(1,403)	(0.021)
	κ-gen	1.593	58,964	0.772
		(0.022)	(624)	(0.027)
NL10	SM	3.391	25,446	1.362
		(0.036)	(318)	(0.038)
	D	4.317	25,653	0.690
		(0.060)	(271)	(0.019)
	κ-gen	3.072	25,010	0.818
		(0.034)	(119)	(0.022)

(continued)

Table B.1 (continued)

Code[a]	Model[b]	Parameters[c]		
		a (α)	b (β)	q, p, κ
NO10	SM	3.139	424,691	2.096
		(0.008)	(1,704)	(0.018)
	D	5.389	398,140	0.470
		(0.017)	(655)	(0.002)
	κ-gen	2.935	354,964	0.598
		(0.007)	(342)	(0.004)
PA13	SM	1.743	6,346	1.420
		(0.018)	(165)	(0.042)
	D	2.109	5,635	0.802
		(0.028)	(121)	(0.022)
	κ-gen	1.560	6,058	0.753
		(0.018)	(57)	(0.023)
PE13	SM	1.331	36,414	4.164
		(0.007)	(997)	(0.116)
	D	2.845	18,481	0.383
		(0.028)	(164)	(0.006)
	κ-gen	1.281	13,419	0.384
		(0.008)	(75)	(0.011)
PL13	SM	2.548	30,263	1.638
		(0.014)	(279)	(0.026)
	D	3.692	29,585	0.582
		(0.023)	(139)	(0.006)
	κ-gen	2.354	27,071	0.727
		(0.015)	(91)	(0.012)
RO97	SM	3.393	7,601,570	1.145
		(0.027)	(71,097)	(0.023)
	D	3.782	7,694,397	0.843
		(0.032)	(59,012)	(0.015)
	κ-gen	3.003	8,009,563	0.894
		(0.021)	(24,993)	(0.015)
RU13	SM	2.502	322,993	1.525
		(0.036)	(8,451)	(0.065)
	D	3.253	303,430	0.706
		(0.068)	(6,211)	(0.029)
	κ-gen	2.255	301,183	0.725
		(0.038)	(2,888)	(0.035)

(continued)

Table B.1 (continued)

Code[a]	Model[b]	Parameters[c]		
		a (α)	b (β)	q, p, κ
RS13	SM	2.040	645,918	2.938
		(0.031)	(28,733)	(0.192)
	D	4.021	499,593	0.405
		(0.087)	(6,786)	(0.013)
	κ-gen	1.937	406,591	0.476
		(0.031)	(4,015)	(0.027)
SK10	SM	3.093	9,568	1.707
		(0.058)	(281)	(0.104)
	D	4.409	9,078	0.614
		(0.098)	(142)	(0.026)
	κ-gen	2.822	8,642	0.676
		(0.049)	(66)	(0.033)
SI10	SM	2.797	20,608	2.972
		(0.050)	(842)	(0.242)
	D	5.324	16,737	0.433
		(0.125)	(193)	(0.016)
	κ-gen	2.646	14,653	0.466
		(0.044)	(115)	(0.030)
ZA12	SM	1.655	17,754	0.860
		(0.021)	(481)	(0.024)
	D	1.439	15,644	1.311
		(0.015)	(347)	(0.028)
	κ-gen	1.497	23,403	1.246
		(0.024)	(351)	(0.037)
KR06	SM	2.226	29,900,000	2.920
		(0.018)	(608,865)	(0.095)
	D	4.167	23,100,000	0.443
		(0.050)	(183,287)	(0.008)
	κ-gen	2.100	19,600,000	0.469
		(0.018)	(96,474)	(0.015)
ES10	SM	1.823	43,627	5.691
		(0.013)	(1,235)	(0.239)
	D	4.038	21,812	0.376
		(0.051)	(182)	(0.007)
	κ-gen	1.778	17,435	0.334
		(0.017)	(105)	(0.017)

(continued)

Table B.1 (continued)

Code[a]	Model[b]	Parameters[c]		
		a (α)	b (β)	q, p, κ
SE05	SM	3.636	216,584	1.473
		(0.037)	(2,670)	(0.044)
	D	4.829	214,737	0.661
		(0.060)	(1,731)	(0.016)
	κ-gen	3.271	209,010	0.735
		(0.031)	(778)	(0.019)
CH04	SM	2.775	64,986	2.190
		(0.057)	(2,573)	(0.165)
	D	4.631	58,195	0.497
		(0.122)	(948)	(0.022)
	κ-gen	2.587	52,105	0.581
		(0.052)	(487)	(0.036)
TW10	SM	2.707	453,027	1.383
		(0.032)	(8,473)	(0.046)
	D	3.336	437,303	0.755
		(0.043)	(5,587)	(0.020)
	κ-gen	2.441	442,313	0.794
		(0.025)	(2,392)	(0.021)
UK10	SM	2.578	19,860	1.369
		(0.020)	(250)	(0.029)
	D	3.315	20,180	0.677
		(0.027)	(154)	(0.010)
	κ-gen	2.340	19,348	0.819
		(0.019)	(85)	(0.016)
US13	SM	2.040	46,735	1.839
		(0.012)	(682)	(0.037)
	D	3.008	41,865	0.592
		(0.021)	(301)	(0.008)
	κ-gen	1.860	38,543	0.625
		(0.010)	(137)	(0.010)
UY13	SM	2.698	191,599	0.932
		(0.017)	(1,688)	(0.014)
	D	2.698	191,599	0.932
		(0.017)	(1,688)	(0.014)
	κ-gen	2.352	224,544	1.062
		(0.016)	(942)	(0.016)

[a]See footnote 28
[b]*SM* = Singh-Maddala, *D* = Dagum type I, κ-gen = κ-generalized
[c]Numbers in parentheses: estimated standard errors
Source Authors' calculations based on LIS

Table B.2 J_α measures of discrepancy: all LIS countries in the most recent years

Code[a]	Model[b]	α	$J_\alpha \times 10^2$	
			Index	Rank
AU10	SM	−1	0.381	1
	D	−1	0.594	2
	κ-gen	−1	0.789	3
	SM	0	0.331	1
	D	0	0.522	2
	κ-gen	0	0.712	3
	SM	1	0.314	1
	D	1	0.487	2
	κ-gen	1	0.669	3
	SM	2	0.311	1
	D	2	0.469	2
	κ-gen	2	0.646	3
AT04	SM	−1	0.150	3
	D	−1	0.113	2
	κ-gen	−1	0.095	1
	SM	0	0.089	3
	D	0	0.073	2
	κ-gen	0	0.053	1
	SM	1	0.071	3
	D	1	0.061	2
	κ-gen	1	0.039	1
	SM	2	0.064	3
	D	2	0.056	2
	κ-gen	2	0.033	1
BE00	SM	−1	0.438	3
	D	−1	0.429	2
	κ-gen	−1	0.340	1
	SM	0	0.494	3
	D	0	0.489	2
	κ-gen	0	0.380	1
	SM	1	0.589	3
	D	1	0.583	2
	κ-gen	1	0.441	1
	SM	2	0.729	3
	D	2	0.722	2
	κ-gen	2	0.525	1

(continued)

Table B.2 (continued)

Code[a]	Model[b]	α	$J_\alpha \times 10^2$	
			Index	Rank
BR13	SM	−1	3.792	2
	D	−1	3.541	1
	κ-gen	−1	13.640	3
	SM	0	2.054	2
	D	0	1.943	1
	κ-gen	0	5.509	3
	SM	1	1.395	2
	D	1	1.331	1
	κ-gen	1	3.314	3
	SM	2	1.072	2
	D	2	1.030	1
	κ-gen	2	2.413	3
CA10	SM	−1	0.162	3
	D	−1	0.090	1
	κ-gen	−1	0.095	2
	SM	0	0.118	3
	D	0	0.065	1
	κ-gen	0	0.066	2
	SM	1	0.105	3
	D	1	0.055	2
	κ-gen	1	0.054	1
	SM	2	0.101	3
	D	2	0.050	2
	κ-gen	2	0.049	1
CN02	SM	−1	0.109	1
	D	−1	2.490	3
	κ-gen	−1	0.151	2
	SM	0	0.107	1
	D	0	1.780	3
	κ-gen	0	0.145	2
	SM	1	0.105	1
	D	1	1.400	3
	κ-gen	1	0.142	2
	SM	2	0.105	1
	D	2	1.181	3
	κ-gen	2	0.142	2

(continued)

Table B.2 (continued)

Code[a]	Model[b]	α	$J_\alpha \times 10^2$	
			Index	Rank
CO10	SM	−1	1.741	1
	D	−1	2.617	2
	κ-gen	−1	7.390	3
	SM	0	1.225	1
	D	0	1.721	2
	κ-gen	0	4.045	3
	SM	1	0.951	1
	D	1	1.282	2
	κ-gen	1	2.723	3
	SM	2	0.788	1
	D	2	1.036	2
	κ-gen	2	2.073	3
CZ10	SM	−1	0.055	2
	D	−1	0.021	1
	κ-gen	−1	0.235	3
	SM	0	0.052	2
	D	0	0.020	1
	κ-gen	0	0.215	3
	SM	1	0.050	2
	D	1	0.019	1
	κ-gen	1	0.199	3
	SM	2	0.048	2
	D	2	0.019	1
	κ-gen	2	0.186	3
DK10	SM	−1	0.877	3
	D	−1	0.618	1
	κ-gen	−1	0.682	2
	SM	0	0.531	3
	D	0	0.421	2
	κ-gen	0	0.408	1
	SM	1	0.509	3
	D	1	0.405	2
	κ-gen	1	0.370	1
	SM	2	0.585	3
	D	2	0.448	2
	κ-gen	2	0.396	1

(continued)

Table B.2 (continued)

Code[a]	Model[b]	α	$J_\alpha \times 10^2$	
			Index	Rank
EG12	SM	−1	2.300	2
	D	−1	2.280	1
	κ-gen	−1	7.255	3
	SM	0	1.580	2
	D	0	1.571	1
	κ-gen	0	4.116	3
	SM	1	1.198	2
	D	1	1.195	1
	κ-gen	1	2.788	3
	SM	2	0.973	1
	D	2	0.974	2
	κ-gen	2	2.115	3
EE10	SM	−1	0.069	1
	D	−1	0.118	2
	κ-gen	−1	0.219	3
	SM	0	0.063	1
	D	0	0.110	2
	κ-gen	0	0.203	3
	SM	1	0.058	1
	D	1	0.105	2
	κ-gen	1	0.190	3
	SM	2	0.055	1
	D	2	0.100	2
	κ-gen	2	0.180	3
FI10	SM	−1	0.135	3
	D	−1	0.095	2
	κ-gen	−1	0.066	1
	SM	0	0.135	3
	D	0	0.094	2
	κ-gen	0	0.063	1
	SM	1	0.145	3
	D	1	0.099	2
	κ-gen	1	0.064	1
	SM	2	0.160	3
	D	2	0.108	2
	κ-gen	2	0.068	1

(continued)

Table B.2 (continued)

Code[a]	Model[b]	α	$J_\alpha \times 10^2$	
			Index	Rank
FR10	SM	−1	0.384	3
	D	−1	0.147	1
	κ-gen	−1	0.245	2
	SM	0	0.354	3
	D	0	0.139	1
	κ-gen	0	0.209	2
	SM	1	0.350	3
	D	1	0.136	1
	κ-gen	1	0.191	2
	SM	2	0.362	3
	D	2	0.136	1
	κ-gen	2	0.182	2
DE10	SM	−1	0.049	2
	D	−1	0.043	1
	κ-gen	−1	0.082	3
	SM	0	0.028	2
	D	0	0.026	1
	κ-gen	0	0.068	3
	SM	1	0.021	2
	D	1	0.020	1
	κ-gen	1	0.062	3
	SM	2	0.018	2
	D	2	0.017	1
	κ-gen	2	0.059	3
GE13	SM	−1	0.007	1
	D	−1	0.009	2
	κ-gen	−1	0.023	3
	SM	0	0.006	1
	D	0	0.007	2
	κ-gen	0	0.017	3
	SM	1	0.005	1
	D	1	0.006	2
	κ-gen	1	0.014	3
	SM	2	0.005	1
	D	2	0.005	2
	κ-gen	2	0.012	3

(continued)

Table B.2 (continued)

Code[a]	Model[b]	α	$J_\alpha \times 10^2$	
			Index	Rank
GR10	SM	−1	0.338	3
	D	−1	0.242	2
	κ-gen	−1	0.215	1
	SM	0	0.235	3
	D	0	0.179	2
	κ-gen	0	0.139	1
	SM	1	0.208	3
	D	1	0.161	2
	κ-gen	1	0.114	1
	SM	2	0.204	3
	D	2	0.158	2
	κ-gen	2	0.105	1
GT06	SM	−1	1.264	1
	D	−1	3.601	2
	κ-gen	−1	3.892	3
	SM	0	0.905	1
	D	0	2.298	2
	κ-gen	0	2.429	3
	SM	1	0.694	1
	D	1	1.656	2
	κ-gen	1	1.727	3
	SM	2	0.562	1
	D	2	1.298	2
	κ-gen	2	1.344	3
HU12	SM	−1	0.202	3
	D	−1	0.160	2
	κ-gen	−1	0.149	1
	SM	0	0.225	3
	D	0	0.177	2
	κ-gen	0	0.158	1
	SM	1	0.267	3
	D	1	0.205	2
	κ-gen	1	0.176	1
	SM	2	0.333	3
	D	2	0.246	2
	κ-gen	2	0.201	1

(continued)

Table B.2 (continued)

Code[a]	Model[b]	α	$J_\alpha \times 10^2$	
			Index	Rank
IS10	SM	−1	0.337	3
	D	−1	0.296	2
	κ-gen	−1	0.239	1
	SM	0	0.330	3
	D	0	0.307	2
	κ-gen	0	0.234	1
	SM	1	0.381	3
	D	1	0.358	2
	κ-gen	1	0.263	1
	SM	2	0.473	3
	D	2	0.446	2
	κ-gen	2	0.315	1
IN04	SM	−1	1.341	1
	D	−1	6.911	3
	κ-gen	−1	6.291	2
	SM	0	1.009	1
	D	0	3.921	3
	κ-gen	0	3.608	2
	SM	1	0.815	1
	D	1	2.707	3
	κ-gen	1	2.507	2
	SM	2	0.694	1
	D	2	2.104	3
	κ-gen	2	1.958	2
IE10	SM	−1	0.220	1
	D	−1	0.226	2
	κ-gen	−1	0.242	3
	SM	0	0.176	1
	D	0	0.196	2
	κ-gen	0	0.217	3
	SM	1	0.156	1
	D	1	0.181	2
	κ-gen	1	0.205	3
	SM	2	0.146	1
	D	2	0.175	2
	κ-gen	2	0.201	3

(continued)

Table B.2 (continued)

Code[a]	Model[b]	α	$J_\alpha \times 10^2$	
			Index	Rank
IL12	SM	−1	0.001	2
	D	−1	0.001	3
	κ-gen	−1	0.001	1
	SM	0	0.001	2
	D	0	0.001	3
	κ-gen	0	0.001	1
	SM	1	0.001	2
	D	1	0.001	3
	κ-gen	1	0.001	1
	SM	2	0.001	2
	D	2	0.001	3
	κ-gen	2	0.001	1
IT10	SM	−1	0.315	3
	D	−1	0.200	2
	κ-gen	−1	0.183	1
	SM	0	0.232	3
	D	0	0.158	2
	κ-gen	0	0.122	1
	SM	1	0.207	3
	D	1	0.142	2
	κ-gen	1	0.098	1
	SM	2	0.201	3
	D	2	0.137	2
	κ-gen	2	0.087	1
JP08	SM	−1	0.126	3
	D	−1	0.078	2
	κ-gen	−1	0.070	1
	SM	0	0.112	3
	D	0	0.072	2
	κ-gen	0	0.059	1
	SM	1	0.106	3
	D	1	0.069	2
	κ-gen	1	0.053	1
	SM	2	0.105	3
	D	2	0.068	2
	κ-gen	2	0.049	1

(continued)

Table B.2 (continued)

Code[a]	Model[b]	α	$J_\alpha \times 10^2$	
			Index	Rank
LU10	SM	−1	0.319	3
	D	−1	0.251	2
	κ-gen	−1	0.247	1
	SM	0	0.163	3
	D	0	0.137	1
	κ-gen	0	0.140	2
	SM	1	0.123	3
	D	1	0.106	1
	κ-gen	1	0.110	2
	SM	2	0.111	3
	D	2	0.096	1
	κ-gen	2	0.100	2
MX12	SM	−1	0.004	1
	D	−1	0.010	2
	κ-gen	−1	0.023	3
	SM	0	0.003	1
	D	0	0.008	2
	κ-gen	0	0.016	3
	SM	1	0.003	1
	D	1	0.006	2
	κ-gen	1	0.013	3
	SM	2	0.003	1
	D	2	0.005	2
	κ-gen	2	0.010	3
NL10	SM	−1	0.314	3
	D	−1	0.294	2
	κ-gen	−1	0.220	1
	SM	0	0.134	2
	D	0	0.147	3
	κ-gen	0	0.090	1
	SM	1	0.101	2
	D	1	0.122	3
	κ-gen	1	0.062	1
	SM	2	0.095	2
	D	2	0.120	3
	κ-gen	2	0.053	1

(continued)

Table B.2 (continued)

Code[a]	Model[b]	α	$J_{\alpha} \times 10^2$	
			Index	Rank
NO10	SM	−1	0.885	3
	D	−1	0.557	1
	κ-gen	−1	0.700	2
	SM	0	0.688	3
	D	0	0.505	1
	κ-gen	0	0.513	2
	SM	1	0.717	3
	D	1	0.550	2
	κ-gen	1	0.503	1
	SM	2	0.855	3
	D	2	0.653	2
	κ-gen	2	0.558	1
PA13	SM	−1	0.285	1
	D	−1	1.393	3
	κ-gen	−1	1.016	2
	SM	0	0.236	1
	D	0	1.048	3
	κ-gen	0	0.771	2
	SM	1	0.202	1
	D	1	0.844	3
	κ-gen	1	0.624	2
	SM	2	0.178	1
	D	2	0.712	3
	κ-gen	2	0.529	2
PE13	SM	−1	0.151	2
	D	−1	0.266	3
	κ-gen	−1	0.086	1
	SM	0	0.154	2
	D	0	0.219	3
	κ-gen	0	0.084	1
	SM	1	0.158	2
	D	1	0.190	3
	κ-gen	1	0.084	1
	SM	2	0.168	2
	D	2	0.206	3
	κ-gen	2	0.092	1

(continued)

Table B.2 (continued)

Code[a]	Model[b]	α	$J_\alpha \times 10^2$	
			Index	Rank
PL13	SM	−1	0.005	3
	D	−1	0.004	1
	κ-gen	−1	0.004	2
	SM	0	0.002	3
	D	0	0.002	2
	κ-gen	0	0.002	1
	SM	1	0.002	3
	D	1	0.002	2
	κ-gen	1	0.001	1
	SM	2	0.002	3
	D	2	0.002	2
	κ-gen	2	0.001	1
RO97	SM	−1	0.059	3
	D	−1	0.048	2
	κ-gen	−1	0.032	1
	SM	0	0.059	3
	D	0	0.048	2
	κ-gen	0	0.031	1
	SM	1	0.061	3
	D	1	0.049	2
	κ-gen	1	0.030	1
	SM	2	0.065	3
	D	2	0.052	2
	κ-gen	2	0.030	1
RU13	SM	−1	0.059	1
	D	−1	0.095	2
	κ-gen	−1	0.169	3
	SM	0	0.051	1
	D	0	0.087	2
	κ-gen	0	0.159	3
	SM	1	0.047	1
	D	1	0.082	2
	κ-gen	1	0.151	3
	SM	2	0.045	1
	D	2	0.078	2
	κ-gen	2	0.145	3

(continued)

Table B.2 (continued)

Code[a]	Model[b]	α	$J_\alpha \times 10^2$	
			Index	Rank
RS13	SM	−1	0.510	3
	D	−1	0.274	1
	κ-gen	−1	0.359	2
	SM	0	0.340	3
	D	0	0.195	1
	κ-gen	0	0.214	2
	SM	1	0.302	3
	D	1	0.171	2
	κ-gen	1	0.171	1
	SM	2	0.302	3
	D	2	0.166	2
	κ-gen	2	0.156	1
SK10	SM	−1	0.215	3
	D	−1	0.186	2
	κ-gen	−1	0.164	1
	SM	0	0.233	3
	D	0	0.200	2
	κ-gen	0	0.169	1
	SM	1	0.272	3
	D	1	0.230	2
	κ-gen	1	0.183	1
	SM	2	0.340	3
	D	2	0.280	2
	κ-gen	2	0.206	1
SI10	SM	−1	0.150	3
	D	−1	0.064	1
	κ-gen	−1	0.103	2
	SM	0	0.129	3
	D	0	0.057	1
	κ-gen	0	0.086	2
	SM	1	0.117	3
	D	1	0.053	1
	κ-gen	1	0.075	2
	SM	2	0.112	3
	D	2	0.050	1
	κ-gen	2	0.068	2

(continued)

Table B.2 (continued)

Code[a]	Model[b]	α	$J_\alpha \times 10^2$	
			Index	Rank
ZA12	SM	−1	13.931	2
	D	−1	12.263	1
	κ-gen	−1	43.116	3
	SM	0	7.531	2
	D	0	6.807	1
	κ-gen	0	18.047	3
	SM	1	5.058	2
	D	1	4.626	1
	κ-gen	1	11.006	3
	SM	2	3.883	2
	D	2	3.567	1
	κ-gen	2	8.286	3
KR06	SM	−1	0.121	3
	D	−1	0.033	1
	κ-gen	−1	0.057	2
	SM	0	0.102	3
	D	0	0.026	1
	κ-gen	0	0.040	2
	SM	1	0.097	3
	D	1	0.023	1
	κ-gen	1	0.033	2
	SM	2	0.097	3
	D	2	0.021	1
	κ-gen	2	0.029	2
ES10	SM	−1	0.280	2
	D	−1	0.288	3
	κ-gen	−1	0.268	1
	SM	0	0.179	1
	D	0	0.230	3
	κ-gen	0	0.181	2
	SM	1	0.144	1
	D	1	0.205	3
	κ-gen	1	0.149	2
	SM	2	0.129	1
	D	2	0.192	3
	κ-gen	2	0.135	2

(continued)

Table B.2 (continued)

Code[a]	Model[b]	α	$J_\alpha \times 10^2$	
			Index	Rank
SE05	SM	−1	0.360	3
	D	−1	0.310	2
	κ-gen	−1	0.269	1
	SM	0	0.358	3
	D	0	0.317	2
	κ-gen	0	0.267	1
	SM	1	0.416	3
	D	1	0.367	2
	κ-gen	1	0.301	1
	SM	2	0.519	3
	D	2	0.450	2
	κ-gen	2	0.356	1
CH04	SM	−1	0.841	3
	D	−1	0.638	1
	κ-gen	−1	0.708	2
	SM	0	0.442	3
	D	0	0.331	2
	κ-gen	0	0.319	1
	SM	1	0.385	3
	D	1	0.276	2
	κ-gen	1	0.240	1
	SM	2	0.414	3
	D	2	0.284	2
	κ-gen	2	0.231	1
TW10	SM	−1	0.015	1
	D	−1	0.044	2
	κ-gen	−1	0.052	3
	SM	0	0.015	1
	D	0	0.043	2
	κ-gen	0	0.051	3
	SM	1	0.015	1
	D	1	0.041	2
	κ-gen	1	0.049	3
	SM	2	0.016	1
	D	2	0.040	2
	κ-gen	2	0.048	3

(continued)

Table B.2 (continued)

Code[a]	Model[b]	α	$J_\alpha \times 10^2$	
			Index	Rank
UK10	SM	−1	0.438	3
	D	−1	0.363	2
	κ-gen	−1	0.340	1
	SM	0	0.265	3
	D	0	0.253	2
	κ-gen	0	0.202	1
	SM	1	0.219	2
	D	1	0.223	3
	κ-gen	1	0.155	1
	SM	2	0.207	2
	D	2	0.217	3
	κ-gen	2	0.135	1
US13	SM	−1	0.096	2
	D	−1	0.105	3
	κ-gen	−1	0.089	1
	SM	0	0.089	2
	D	0	0.090	3
	κ-gen	0	0.077	1
	SM	1	0.087	3
	D	1	0.079	2
	κ-gen	1	0.068	1
	SM	2	0.087	3
	D	2	0.072	2
	κ-gen	2	0.063	1
UY13	SM	−1	0.995	1
	D	−1	1.154	2
	κ-gen	−1	2.573	3
	SM	0	0.728	1
	D	0	0.838	2
	κ-gen	0	1.698	3
	SM	1	0.575	1
	D	1	0.660	2
	κ-gen	1	1.272	3
	SM	2	0.479	1
	D	2	0.549	2
	κ-gen	2	1.034	3

[a]See footnote 28

[b]*SM* = Singh-Maddala, *D* = Dagum type I, κ-gen = κ-generalized

Source Authors' calculations based on LIS

Table B.3 Selected distributional statistics implied by the Singh-Maddala, Dagum type I and κ-generalized parameter estimates: all LIS countries in the most recent years[a]

Code[b]	Model[c]	Mean	G	MLD	GE (2)[d]	A (1)	S
AU10	SM	49,235	0.332	0.196	0.232	0.178	11.573
	D	49,448	0.335	0.203	0.267	0.184	11.578
	κ-gen	49,407	0.334	0.203	0.267	0.184	9.577
AT04	SM	22,495	0.269	0.123	0.153	0.116	10.605
	D	22,406	0.267	0.123	0.152	0.116	10.602
	κ-gen	22,586	0.273	0.129	0.181	0.121	8.608
BE00	SM	741,498	0.273	0.128	0.159	0.120	14.116
	D	740,563	0.273	0.129	0.161	0.121	14.117
	κ-gen	742,666	0.274	0.132	0.173	0.123	12.120
BR13	SM	19,567	0.462	0.383	1.576	0.318	10.749
	D	19,667	0.464	0.388	2.077	0.322	10.750
	κ-gen	20,181	0.479	0.417	–	0.341	8.752
CA10	SM	41,746	0.316	0.177	0.205	0.162	11.373
	D	41,784	0.317	0.182	0.223	0.166	11.375
	κ-gen	41,852	0.318	0.182	0.230	0.166	9.376
CN02	SM	7,904	0.513	0.532	0.880	0.412	9.939
	D	8,953	0.572	0.638	–	0.471	9.969
	κ-gen	7,901	0.509	0.541	0.968	0.418	7.945
CO10	SM	7,251,501	0.496	0.461	–	0.369	16.706
	D	7,348,717	0.504	0.477	4.989	0.380	16.706
	κ-gen	7,596,673	0.521	0.511	1.374	0.400	14.709
CZ10	SM	245,763	0.257	0.111	0.144	0.105	12.946
	D	245,553	0.256	0.111	0.144	0.105	12.945
	κ-gen	246,710	0.261	0.117	0.169	0.110	10.952
DK10	SM	244,744	0.250	0.109	0.111	0.103	12.964
	D	243,277	0.248	0.113	0.112	0.106	12.963
	κ-gen	244,736	0.250	0.111	0.116	0.105	10.965
EG12	SM	10,855	0.476	0.411	2.095	0.337	10.167
	D	10,811	0.475	0.412	2.468	0.338	10.163
	κ-gen	11,310	0.499	0.455	–	0.366	8.169
EE10	SM	116,895	0.327	0.192	0.216	0.174	12.434
	D	118,100	0.334	0.201	0.267	0.182	12.444
	κ-gen	117,348	0.330	0.199	0.248	0.181	10.440
FI10	SM	26,339	0.258	0.114	0.134	0.108	10.739
	D	26,424	0.260	0.116	0.143	0.110	10.743
	κ-gen	26,372	0.259	0.118	0.144	0.111	8.744

(continued)

Table B.3 (continued)

Code[b]	Model[c]	Mean	G	MLD	GE (2)[d]	A (1)	S
FR10	SM	23,718	0.288	0.143	0.173	0.133	10.724
	D	23,565	0.286	0.145	0.174	0.135	10.720
	κ-gen	23,801	0.291	0.149	0.203	0.138	8.726
DE10	SM	22,799	0.285	0.139	0.176	0.130	10.668
	D	22,829	0.286	0.141	0.185	0.132	10.670
	κ-gen	22,873	0.288	0.145	0.201	0.135	8.672
GE13	SM	4,574	0.407	0.292	0.665	0.254	9.260
	D	4,684	0.420	0.308	1.032	0.265	9.269
	κ-gen	4,646	0.417	0.312	1.309	0.268	7.269
GR10	SM	15,030	0.334	0.201	0.230	0.182	10.398
	D	14,954	0.333	0.207	0.245	0.187	10.395
	κ-gen	15,072	0.336	0.207	0.264	0.187	8.399
GT06	SM	20,706	0.506	0.482	1.809	0.382	10.851
	D	21,870	0.531	0.527	–	0.409	10.866
	κ-gen	21,229	0.517	0.512	–	0.401	8.857
HU12	SM	1,628,570	0.288	0.146	0.159	0.136	14.972
	D	1,633,128	0.290	0.150	0.174	0.140	14.976
	κ-gen	1,632,122	0.290	0.150	0.175	0.140	12.975
IS10	SM	3,855,452	0.241	0.099	0.112	0.094	15.669
	D	3,838,006	0.240	0.100	0.112	0.095	15.667
	κ-gen	3,859,830	0.242	0.102	0.123	0.097	13.672
IN04	SM	21,051	0.503	0.493	1.006	0.389	10.900
	D	21,708	0.521	0.530	5.336	0.411	10.901
	κ-gen	22,067	0.529	0.545	–	0.420	8.905
IE10	SM	25,332	0.297	0.155	0.171	0.144	10.831
	D	25,477	0.301	0.162	0.197	0.149	10.839
	κ-gen	25,400	0.299	0.161	0.191	0.149	8.838
IL12	SM	94,341	0.372	0.259	0.282	0.228	12.318
	D	96,050	0.381	0.275	0.379	0.240	12.333
	κ-gen	94,562	0.372	0.265	0.304	0.233	10.322
IT10	SM	17,462	0.330	0.196	0.220	0.178	10.542
	D	17,395	0.330	0.202	0.239	0.183	10.539
	κ-gen	17,521	0.333	0.202	0.255	0.183	8.543
JP08	SM	3,315,331	0.302	0.159	0.186	0.147	15.709
	D	3,307,061	0.301	0.162	0.196	0.150	15.707
	κ-gen	3,326,309	0.304	0.165	0.215	0.152	13.711

(continued)

Table B.3 (continued)

Code[b]	Model[c]	Mean	G	MLD	$GE\,(2)$[d]	$A\,(1)$	S
LU10	SM	41,098	0.270	0.125	0.150	0.118	11.221
	D	41,255	0.273	0.129	0.163	0.121	11.226
	κ-gen	41,204	0.274	0.131	0.169	0.123	9.230
MX12	SM	68,769	0.466	0.405	0.850	0.333	12.056
	D	69,101	0.470	0.418	1.368	0.342	12.052
	κ-gen	70,883	0.483	0.439	5.924	0.355	10.058
NL10	SM	24,794	0.259	0.114	0.132	0.108	10.684
	D	24,674	0.258	0.117	0.134	0.110	10.683
	κ-gen	24,865	0.263	0.120	0.153	0.113	8.689
NO10	SM	336,744	0.246	0.106	0.106	0.100	13.274
	D	333,415	0.243	0.109	0.105	0.103	13.264
	κ-gen	336,802	0.246	0.107	0.112	0.102	11.272
PA13	SM	7,117	0.476	0.423	1.090	0.345	9.782
	D	7,374	0.493	0.451	3.842	0.363	9.793
	κ-gen	7,266	0.486	0.447	5.252	0.360	7.786
PE13	SM	13,670	0.459	0.432	0.457	0.351	10.498
	D	13,777	0.459	0.450	0.621	0.362	10.494
	κ-gen	13,723	0.460	0.437	0.528	0.354	8.496
PL13	SM	27,131	0.319	0.179	0.213	0.164	10.946
	D	26,877	0.317	0.183	0.220	0.167	10.939
	κ-gen	27,262	0.323	0.187	0.261	0.170	8.947
RO97	SM	8,126,608	0.277	0.130	0.169	0.122	16.513
	D	8,117,316	0.276	0.131	0.171	0.123	16.513
	κ-gen	8,153,451	0.280	0.136	0.195	0.127	14.517
RU13	SM	304,145	0.332	0.194	0.245	0.177	13.384
	D	306,072	0.336	0.201	0.286	0.182	13.389
	κ-gen	305,781	0.336	0.203	0.298	0.184	11.390
RS13	SM	388,241	0.339	0.214	0.214	0.193	13.680
	D	384,818	0.336	0.223	0.224	0.200	13.670
	κ-gen	388,550	0.339	0.216	0.231	0.195	11.677
SK10	SM	8,351	0.263	0.120	0.128	0.113	9.624
	D	8,349	0.264	0.123	0.136	0.116	9.626
	κ-gen	8,366	0.265	0.124	0.141	0.117	7.628
SI10	SM	13,605	0.256	0.117	0.110	0.111	10.109
	D	13,584	0.255	0.122	0.115	0.115	10.109
	κ-gen	13,609	0.255	0.119	0.114	0.112	8.108

(continued)

Table B.3 (continued)

Code[b]	Model[c]	Mean	G	MLD	GE (2)[d]	A (1)	S
ZA12	SM	50,961	0.677	0.899	–	0.593	11.493
	D	51,503	0.679	0.899	–	0.593	11.499
	κ-gen	72,191	0.774	1.250	–	0.713	9.499
KR06	SM	18,517,582	0.315	0.182	0.178	0.166	17.491
	D	18,530,248	0.314	0.189	0.194	0.172	17.492
	κ-gen	18,538,458	0.315	0.185	0.189	0.169	15.491
ES10	SM	16,173	0.344	0.229	0.206	0.204	10.525
	D	16,201	0.346	0.241	0.235	0.214	10.526
	κ-gen	16,184	0.344	0.230	0.215	0.205	8.524
SE05	SM	202,526	0.236	0.095	0.103	0.090	12.713
	D	202,200	0.236	0.097	0.105	0.092	12.714
	κ-gen	202,722	0.237	0.098	0.110	0.093	10.716
CH04	SM	49,603	0.273	0.132	0.133	0.124	11.448
	D	49,322	0.273	0.137	0.140	0.128	11.446
	κ-gen	49,648	0.274	0.135	0.144	0.127	9.449
TW10	SM	450,171	0.319	0.178	0.230	0.163	13.741
	D	452,099	0.322	0.182	0.257	0.167	13.744
	κ-gen	452,911	0.324	0.186	0.285	0.170	11.746
UK10	SM	20,051	0.336	0.197	0.268	0.179	10.662
	D	19,895	0.334	0.200	0.276	0.181	10.657
	κ-gen	20,209	0.343	0.209	0.361	0.188	8.666
US13	SM	39,160	0.377	0.260	0.328	0.229	11.428
	D	39,417	0.381	0.270	0.402	0.237	11.432
	κ-gen	39,369	0.380	0.269	0.401	0.236	9.430
UY13	SM	258,733	0.386	0.256	0.664	0.226	13.240
	D	261,640	0.391	0.262	0.749	0.230	13.244
	κ-gen	263,070	0.397	0.276	1.396	0.241	11.249

[a] G = Gini, MLD = mean logarithmic deviation, GE (2) = top-sensitive generalized entropy index, A (1) = Atkinson index with inequality aversion parameter $\epsilon = 1$, S = Shannon entropy
[b] See footnote 28
[c] SM = Singh-Maddala, D = Dagum type I, κ-gen = κ-generalized
[d] Cases for which calculation has not been possible are indicated by a dash ("–") symbol—see footnote 33
Source Authors' calculations based on LIS

Table B.4 Estimated mixture models of net wealth distribution: all LWS countries in the most recent years

Code[a]	Model[b]	Parameters[c]						
		s	λ	θ_1	θ_2	a (α)	b (β)	q, p, κ
CA99w	SM	0.859	6,615	0.186	0.022	0.715	7,795,561	30.442
		(0.012)	(149)	(0.003)	(0.001)	(0.007)	(3,794,552)	(9.625)
	D	0.859	6,615	0.186	0.022	2.206	144,091	0.257
		(0.012)	(149)	(0.003)	(0.001)	(0.039)	(2,790)	(0.006)
	κ-gen	0.859	6,615	0.186	0.022	0.726	66,156	0.190
		(0.012)	(149)	(0.003)	(0.001)	(0.006)	(899)	(0.016)
CY02w	SM	0.651	7,651	0.071	0.025	1.006	703,845	19.284
		(0.099)	(2,503)	(0.014)	(0.008)	(0.052)	(491,270)	(11.678)
	D	0.651	7,651	0.071	0.025	3.512	70,370	0.212
		(0.099)	(2,503)	(0.014)	(0.008)	(0.377)	(5,288)	(0.030)
	κ-gen	0.651	7,651	0.071	0.025	1.001	38,051	0.143
		(0.099)	(2,503)	(0.014)	(0.008)	(0.049)	(2,267)	(0.070)
FI98w	SM	0.701	7,660	0.142	0.013	0.828	3,068,647	29.780
		(0.021)	(492)	(0.006)	(0.002)	(0.015)	(1,983,863)	(14.320)
	D	0.701	7,660	0.142	0.013	2.933	105,525	0.217
		(0.021)	(492)	(0.006)	(0.002)	(0.099)	(2,820)	(0.010)
	κ-gen	0.701	7,660	0.142	0.013	0.849	50,960	0.201
		(0.021)	(492)	(0.006)	(0.002)	(0.014)	(1,137)	(0.025)
IT04w	SM	0.897	4,143	0.040	0.060	0.848	5,498,769	27.897
		(0.036)	(275)	(0.002)	(0.003)	(0.011)	(2,487,096)	(9.642)
	D	0.897	4,143	0.040	0.060	2.799	220,474	0.236
		(0.036)	(275)	(0.002)	(0.003)	(0.066)	(4,350)	(0.008)
	κ-gen	0.897	4,143	0.040	0.060	0.858	109,377	0.182
		(0.036)	(275)	(0.002)	(0.003)	(0.009)	(1,647)	(0.019)
JP03w	SM	0.730	2,893,727	0.119	0.065	0.929	128,000,000	7.578
		(0.027)	(229,462)	(0.006)	(0.005)	(0.023)	(42,600,000)	(1.881)
	D	0.730	2,893,745	0.119	0.065	2.324	25,900,000	0.324
		(0.027)	(229,464)	(0.006)	(0.005)	(0.098)	(1,170,336)	(0.020)
	κ-gen	0.730	2,893,745	0.119	0.065	0.910	15,300,000	0.271
		(0.027)	(229,464)	(0.006)	(0.005)	(0.019)	(412,338)	(0.038)
NO02w	SM	1.001	346,550	0.557	0.002	0.907	1,099,086	3.151
		(0.007)	(3,224)	(0.003)	(0.000)	(0.011)	(87,135)	(0.163)
	D	1.001	346,550	0.557	0.002	1.927	599,083	0.371
		(0.007)	(3,224)	(0.003)	(0.000)	(0.034)	(13,860)	(0.010)
	κ-gen	1.001	346,550	0.557	0.002	0.867	356,942	0.452
		(0.007)	(3,224)	(0.003)	(0.000)	(0.009)	(5,050)	(0.017)

(continued)

Table B.4 (continued)

Code[a]	Model[b]	Parameters[c]						
		s	λ	θ_1	θ_2	a (α)	b (β)	q, p, κ
SE02w	SM	0.733	114,028	0.273	0.031	0.922	2,932,812	5.397
		(0.007)	(2,346)	(0.003)	(0.001)	(0.010)	(318,125)	(0.415)
	D	0.733	114,029	0.273	0.031	2.154	869,919	0.343
		(0.007)	(2,346)	(0.003)	(0.001)	(0.041)	(19,477)	(0.010)
	κ-gen	0.733	114,029	0.273	0.031	0.887	517,736	0.301
		(0.007)	(2,346)	(0.003)	(0.001)	(0.008)	(6,157)	(0.015)
UK00w[d]	SM	–	–	–	–	–	–	–
		–	–	–	–	–	–	–
	D	0.585	1,965	0.127	0.050	2.385	116,786	0.250
		(0.018)	(155)	(0.005)	(0.003)	(0.084)	(4,182)	(0.012)
	κ-gen	0.585	1,965	0.127	0.050	0.759	55,992	0.057
		(0.018)	(155)	(0.005)	(0.003)	(0.012)	(1,365)	(0.074)
USS06w	SM	0.777	10,643	0.216	0.036	0.683	669,965	3.708
		(0.008)	(209)	(0.003)	(0.001)	(0.007)	(68,100)	(0.188)
	D	0.777	10,643	0.216	0.036	1.453	216,638	0.382
		(0.008)	(209)	(0.003)	(0.001)	(0.020)	(5,354)	(0.008)
	κ-gen	0.777	10,643	0.216	0.036	0.660	114,248	0.426
		(0.008)	(209)	(0.003)	(0.001)	(0.005)	(1,683)	(0.014)

[a]See Table 4.1

[b]*SM* = Singh-Maddala mixture model, *D* = Dagum type I mixture model, κ-gen = κ-generalized mixture model

[c]Numbers in parentheses: estimated standard errors

[d]Estimation results for the Singh-Maddala mixture model of net wealth distribution are not reported due to non-convergence of the maximum likelihood algorithm

Source Authors' calculations based on LWS

Table B.5 Density, distribution function and quantile function of the inverses of the extended κ-generalized distributions of the first and second kind (IEκG1 and IEκG2)

Model	$f(x), x > 0$	$F(x), x > 0$	$F^{-1}(u), 0 < u < 1$
IEκG1 (a, b, p, r)[a]	$\dfrac{a\left[-u^r \frac{u^{\frac{1}{2p}} - u^{-\frac{1}{2p}}}{1/p}\right]^{\frac{1}{a}+1}}{b\left[\left(pr+\frac{1}{2}\right)u^{r+\frac{1}{2p}-1} - \left(pr-\frac{1}{2}\right)u^{r-\frac{1}{2p}-1}\right]}$	–	$b\left[-u^r \frac{u^{\frac{1}{2p}} - u^{-\frac{1}{2p}}}{1/p}\right]^{-\frac{1}{a}}$
IEκG2 (a, b, p, q)	$\frac{a}{bB(p,q)} \frac{z^{p-\frac{1}{2a}}(1-z)^{q+\frac{1}{a}}}{2(1+z)}$, $z = 1 - \left(\frac{x}{b}\right)^{-a}\left[\sqrt{1 + \frac{1}{4}\left(\frac{x}{b}\right)^{-2a}} - \frac{1}{2}\left(\frac{x}{b}\right)^{-a}\right]$	$I_z(p, q)$, $z = 1 - \left(\frac{x}{b}\right)^{-a}\left[\sqrt{1 + \frac{1}{4}\left(\frac{x}{b}\right)^{-2a}} - \frac{1}{2}\left(\frac{x}{b}\right)^{-a}\right]$	$bz^{\frac{1}{2a}}(1-z)^{-\frac{1}{a}}$, $z = I_u^{-1}(p, q)$

[a]Similarly to the EκG1 distribution, the probability density function of the IEκG1 is expressed in terms of the cumulative probabilities u; furthermore, no closed-form expression of the cumulative distribution function is available in this case

Source Okamoto (2013)

Table B.6 Moments and income inequality tools for the four-parameter variants of the κ-generalized distribution

Model	Moment[a]	Lorenz curve	Gini index[b]
EκG1 (a, b, q, r)	$b^h q^{\frac{h}{a}+1} B\left(1+\frac{h}{a}, q-\frac{h}{2a}+\frac{hqr}{a}\right)$	$I_z\left(1+\frac{1}{a}, q-\frac{1}{2a}+\frac{2qr}{a}\right)$, $z = 1-(1-u)^{\frac{1}{q}}$	$1-2\frac{B\left(1+\frac{1}{a}, 2q-\frac{1}{2a}+\frac{qr}{a}\right)}{B\left(1+\frac{1}{a}, q-\frac{1}{2a}+\frac{qr}{a}\right)}$
EκG2 (a, b, p, q)	$b^h \frac{B\left(p+\frac{h}{a}, q-\frac{h}{2a}\right)}{B(p,q)}$	$I_z\left(p+\frac{1}{a}, q-\frac{1}{2a}\right)$, $z = I_u^{-1}(p, q)$	$\frac{B\left(2p+\frac{1}{a}, 2q-\frac{1}{2a}\right)}{B(p,q)B\left(p+\frac{1}{a}, q-\frac{1}{2a}\right)}\left[\frac{1}{p}\,{}_3F_2\left(1, p+q, 2p+\frac{1}{a}; p+1, 2p+2q+\frac{1}{2a}; 1\right)\right.$ $\left.-\frac{1}{p+\frac{1}{a}}\,{}_3F_2\left(1, p+q+\frac{1}{2a}, 2p+\frac{1}{a}; p+\frac{1}{a}+1, 2p+2q+\frac{1}{2a}; 1\right)\right]$
IEκG1 (a, b, p, r)	$b^h p^{\frac{h}{a}-1} B\left(p+\frac{h}{2a}-\frac{hpr}{a}, 1-\frac{h}{a}\right)$	$I_z\left(p+\frac{1}{2a}-\frac{2pr}{a}, 1-\frac{1}{a}\right)$, $z = u^{\frac{1}{p}}$	$2\frac{B\left(2p+\frac{1}{2a}-\frac{pr}{a}, 1-\frac{1}{a}\right)}{B\left(p+\frac{1}{2a}-\frac{pr}{a}, 1-\frac{1}{a}\right)}-1$
IEκG2 (a, b, p, q)	$b^h \frac{B\left(p+\frac{h}{2a}, q-\frac{h}{a}\right)}{B(p,q)}$	$I_z\left(p+\frac{1}{2a}, q-\frac{1}{a}\right)$, $z = I_u^{-1}(p, q)$	$\frac{B\left(2p+\frac{1}{2a}, 2q-\frac{1}{a}\right)}{B(p,q)B\left(p+\frac{1}{2a}, q-\frac{1}{a}\right)}\left[\frac{1}{p}\,{}_3F_2\left(1, p+q, 2p+\frac{1}{2a}; p+1, 2p+2q-\frac{1}{2a}; 1\right)\right.$ $\left.-\frac{1}{p+\frac{1}{2a}}\,{}_3F_2\left(1, p+q-\frac{1}{2a}, 2p+\frac{1}{2a}; p+\frac{1}{2a}+1, 2p+2q-\frac{1}{2a}; 1\right)\right]$

Model	Coefficient of variation	Mean logarithmic deviation	Theil index
EκG1 (a, b, q, r)	$\sqrt{\frac{B\left(1+\frac{2}{a}, q-\frac{1}{a}+\frac{2qr}{a}\right)}{qB\left(1+\frac{1}{a}, q-\frac{1}{2a}+\frac{qr}{a}\right)^2}-1}$	$\ln\left[B\left(1+\frac{1}{a}, q-\frac{1}{2a}+\frac{qr}{a}\right)\right]+\ln(q)-\frac{1}{a}\psi(1)$ $+\left(\frac{1}{2a}-\frac{qr}{a}\right)\psi(q)+\left(\frac{1}{2a}+\frac{qr}{a}\right)\psi(1+q)$	$\frac{1}{a}\psi\left(1+\frac{1}{a}\right)-\left(\frac{1}{2a}-\frac{qr}{a}\right)\psi\left(q-\frac{1}{2a}+\frac{qr}{a}\right)$ $-\left(\frac{1}{2a}+\frac{qr}{a}\right)\psi\left(1+q+\frac{1}{2a}+\frac{qr}{a}\right)$ $-\ln\left[B\left(1+\frac{1}{a}, q-\frac{1}{2a}+\frac{qr}{a}\right)\right]-\ln(q)$

(continued)

Table B.6 (continued)

Model	Coefficient of Variation	Mean logarithmic deviation	Theil index
EκG2 (a, b, p, q)	$\sqrt{\frac{B(p,q)B\left(p+\frac{2}{a},q-\frac{1}{a}\right)}{B\left(p+\frac{1}{a},q-\frac{1}{2a}\right)^2}-1}$	$\ln\left[B\left(p+\frac{1}{a},q-\frac{1}{2a}\right)\right]-\ln\left[B\left(p,q\right)\right]-\frac{1}{a}\psi\left(p\right)$ $+\frac{1}{2a}\psi\left(q\right)+\frac{1}{2a}\psi\left(p+q\right)$	$\frac{1}{a}\psi\left(p+\frac{1}{a}\right)-\frac{1}{2a}\psi\left(q-\frac{1}{2a}\right)-\frac{1}{2a}\psi\left(p+q+\frac{1}{2a}\right)$ $-\ln\left[B\left(p+\frac{1}{a},q-\frac{1}{2a}\right)\right]+\ln\left[B\left(p,q\right)\right]$
IEκG1 (a, b, p, r)	$\sqrt{\frac{pB\left(p+\frac{1}{a}+\frac{2pr}{a},1-\frac{2}{a}\right)}{B\left(p+\frac{1}{2a}-\frac{pr}{a},1-\frac{1}{a}\right)^2}-1}$	$\ln\left[B\left(p+\frac{1}{2a}-\frac{pr}{a},1-\frac{1}{a}\right)\right]+\ln\left(p\right)+\frac{1}{a}\psi\left(1\right)$ $-\left(\frac{1}{2a}-\frac{pr}{a}\right)\psi\left(p\right)-\left(\frac{1}{2a}+\frac{pr}{a}\right)\psi\left(1+p\right)$	$-\frac{1}{a}\psi\left(1-\frac{1}{a}\right)+\left(\frac{1}{2a}-\frac{pr}{a}\right)\psi\left(p+\frac{1}{2a}-\frac{pr}{a}\right)$ $+\left(\frac{1}{2a}+\frac{pr}{a}\right)\psi\left(1+p-\frac{1}{2a}-\frac{pr}{a}\right)$ $-\ln\left[B\left(p+\frac{1}{2a}-\frac{pr}{a},1-\frac{1}{a}\right)\right]+\ln\left(p\right)$
IEκG2 (a, b, p, q)	$\sqrt{\frac{B(p,q)B\left(p+\frac{2}{a},q-\frac{2}{a}\right)}{B\left(p+\frac{1}{2a},q-\frac{1}{a}\right)^2}-1}$	$\ln\left[B\left(p+\frac{1}{2a},q-\frac{1}{a}\right)\right]-\ln\left[B\left(p,q\right)\right]$ $-\frac{1}{2a}\psi\left(p\right)+\frac{1}{a}\psi\left(q\right)-\frac{1}{2a}\psi\left(p+q\right)$	$\frac{1}{2a}\psi\left(p+\frac{1}{2a}\right)-\frac{1}{a}\psi\left(q-\frac{1}{a}\right)$ $+\frac{1}{2a}\psi\left(p+q-\frac{1}{2a}\right)-\ln\left[B\left(p+\frac{1}{2a},q-\frac{1}{a}\right)\right]$ $+\ln\left[B\left(p,q\right)\right]$

[a] $h > 0$ denotes the order of the moment

[b] ${}_3F_2\left(\theta_1, \theta_2, \theta_3; \theta_4, \theta_5; 1\right)$ is the generalized hypergeometric function—see http://mathworld.wolfram.com/GeneralizedHypergeometricFunction.html

Source Okamoto (2013)

References

Abul-Magd, A.Y.: Wealth distribution in an ancient Egyptian society. Phys. Rev. E **66**, 057104 (2002)

Aitchison, J., Brown, J.A.C.: On criteria for descriptions of income distribuition. Metroeconomica **6**, 88–107 (1954)

Aitchison, J., Brown, J.A.C.: The Lognormal Distribution with Special Reference to its Use in Economics. Cambridge University Press, New York (1957)

Allison, P.D.: Measures of inequality. Am. Sociol. Rev. **43**, 865–880 (1978)

Amiel, Y., Cowell, F.A., Polovin, A.: Inequality among the Kibbutzim. Economica **63**, S63–S85 (1996)

Anderson, P.W.: More is different. Science **177**, 393–396 (1972)

Aoki, M.: New Approaches to Macroeconomic Modeling: Evolutionary Stochastic Dynamics, Multiple Equilibria, and Externalities as Field Effects. Cambridge University Press, Cambridge (1996)

Aoki, M.: Modeling Aggregate Behavior and Fluctuations in Economics: Stochastic Views of Interacting Agents. Cambridge University Press, Cambridge (2002)

Aoki, M., Yoshikawa, H.: Reconstructing Macroeconomics: A Perspective from Statistical Physics and Combinatorial Stochastic Processes. Cambridge University Press, Cambridge (2007)

Araar, A., Duclos, J.Y.: User Manual for Stata Package DASP: Version 2.3. PEP, World Bank, UNDP and Université Laval (2013). http://dasp.ecn.ulaval.ca/modules/DASP_V2.3/DASP_MANUAL_V2.3.pdf

Arnold, B.C.: Majorization and the Lorenz Order: A Brief Introduction. Springer, Berlin (1987)

Arnold, B.C., Laguna, L.: On Generalized Pareto Distributions with Applications to Income Data. Iowa State University Press, Ames (1977)

Atkinson, A.B.: On the measurement of inequality. J. Econ. Theory **2**, 244–263 (1970)

Atkinson, A.B.: The distribution of wealth in Britain in the 1960–the estate duty method rexamined. In: Smith, J.D. (ed.) The Personal Distribution of Income and Wealth, pp. 277–319. Columbia University Press, New York (1975)

Atkinson, A.B.: Bringing income distribution in from the cold. Econ. J. **107**, 297–321 (1997)

Atkinson, A.B.: Inequality: What Can Be Done? Harvard University Press, Cambridge (2015)

Atkinson, A.B., Stiglitz, J.E.: Lectures on Public Economics. Princeton University Press, Princeton and Oxford (2015)

Atkinson, A.B., Rainwater, L., Smeeding, T.M.: Income Distribution in OECD Countries: Evidence from Luxembourg Income Study. Organization for Economic Co-operation and Development, Paris (1995)

F. Clementi and M. Gallegati, *The Distribution of Income and Wealth*,
New Economic Windows, DOI 10.1007/978-3-319-27410-2

Atoda, N., Suruga, T., Tachibanaki, T.: Statistical inference of functional forms for income distribution. Econ. Stud. Q. **39**, 14–40 (1988)

Bak, P.: How Nature Works: The Science of Self-Organised Criticality. Copernicus Press, New York (1996)

Bandourian, R., McDonald, J.B., Turley, R.S.: Income distributions: an intertemporal comparison over countries. Estadistica **55**, 135–152 (2003)

Bandyopadhyay, S., Cowell, F.A., Flachaire, E.: Goodness-of-Fit: An Economic Approach. Discussion Paper 444, Department of Economics, University of Oxford, Oxford, UK (2009). http://www.economics.ox.ac.uk/Department-of-Economics-Discussion-Paper-Series/goodness-of-fit-an-economic-approach

Bartels, C.P.A.: Economic Aspects of Regional Welfare: Income Distribution and Unemployment. Martinus Nijhoff, Leiden (1977)

Bartels, C.P.A., van Metelen, H.: Alternative Probability Density Functions of Income: A Comparison of the Longnormal-, Gamma- and Weibull-distribution with Dutch Data. Research Memorandum 29, Vrije Universiteit, Faculty of Economics, Department of Quantitative Studies, Amsterdam (1975)

Beirlant, J., Goegebeur, Y., Teugels, J., Segers, J.: Statistics of Extremes: Theory and Applications. Wiley, West Sussex (2004)

Betti, G., D'Agostino, A., Lemmi, A.: Fuzzy monetary poverty measures under a Dagum income distributive hypothesis. In: Chotikapanich, D. (ed.) Modeling Income Distributions and Lorenz Curves, pp. 225–240. Springer, New York (2008)

Biewen, M., Jenkins, S.P.: A framework for the decomposition of poverty differences with an application to poverty differences between countries. Empir. Econ. **30**, 331–358 (2005)

Bordley, R.F., McDonald, J.B., Mantrala, A.: Something new, something old: parametric models for the size distribution of income. J. Income Distrib. **6**, 91–103 (1996)

Brachmann, K., Stich, A., Trede, M.: Evaluating parametric income distribution models. Allg. Stat. Arch. **80**, 285–298 (1996)

Brandolini, A.: The distribution of wealth in Germany and Sweden: discussion of the papers by Stein and Klevmarken. In: Chaloupek, G., Zotter, T. (eds.) Steigende wirtschaftliche Ungleichheit bei steigendem Reichtum? Tagung der Kammer für Arbeiter und Angestellte für Wien, pp. 45–54. LexisNexis Verlag ARD Orac, Vienna (2006)

Burkhauser, R.V., Feng, S., Jenkins, S.P., Larrimore, J.: Estimating trends in US income inequality using the Current Population Survey: the importance of controlling for censoring. J. Econ. Inequal. **9**, 393–415 (2011)

Canberra Group: Handbook on Household Income Statistics, 2nd edn. United Nations, New York and Geneva (2011)

Capasso, M., Alessi, L., Barigozzi, M., Fagiolo, G.: On approximating the distributions of goodness-of-fit test statistics based on the empirical distribution function: the case of unknown parameters. Adv. Complex Syst. **12**, 157–167 (2009)

Caticha, A.: Entropic Inference and the Foundations of Physics. USB Press, São Paulo (2012). http://www.albany.edu/physics/ACaticha-EIFP-book.pdf

Chakravarty, S.R.: Inequality, Polarization and Poverty: Advances in Distributional Analysis. Springer, New York (2009)

Champernowne, D.G.: A model of income distribution. Econ. J. **63**, 318–351 (1953)

Champernowne, D.G., Cowell, F.A.: Economic Inequality and Income Distribution. Cambridge University Press, Cambridge (1998)

Chandra, M., Singpurwalla, N.D.: Relationships between some notions which are common to reliability theory and economics. Math. Oper. Res. **6**, 113–121 (1981)

Chen, C.-N., Tsaur, T.-W., Rhai, T.-S.: The Gini coefficient and negative income. Oxford Econ. Papers **34**, 473–478 (1982)

Chotikapanich, D. (ed.): Modeling Income Distributions and Lorenz Curves. Springer, New York (2008)

Cingano, F.: Trends in Income Inequality and its Impact on Economic Growth. OECD Social, Employment and Migration Working Paper 163, Organisation for Economic Co-operation and Development, Paris (2014). http://www.oecd-ilibrary.org/social-issues-migration-health/trends-in-income-inequality-and-its-impact-on-economic-growth_5jxrjncwxv6j-en

Clementi, F., Gallegati, M., Kaniadakis, G.: κ-generalized statistics in personal income distribution. Eur. Phys. J. B **57**, 187–193 (2007)

Clementi, F., Di Matteo, T., Gallegati, M., Kaniadakis, G.: The κ-generalized distribution: a new descriptive model for the size distribution of incomes. Phys. A: Stat. Mech. Appl. **387**, 3201–3208 (2008)

Clementi, F., Gallegati, M., Kaniadakis, G.: A κ-generalized statistical mechanics approach to income analysis. J. Stat. Mech.: Theory Exp. **2009**, P02037 (2009)

Clementi, F., Gallegati, M., Kaniadakis, G.: A model of personal income distribution with application to Italian data. Empir. Econ. **39**, 559–591 (2010)

Clementi, F., Gallegati, M., Kaniadakis, G.: A new model of income distribution: the κ-generalized distribution. J. Econ. **105**, 63–91 (2012a)

Clementi, F., Gallegati, M., Kaniadakis, G.: A generalized statistical model for the size distribution of wealth. J. Stat. Mech. Theory Exp. **2012**, P12006 (2012b)

Cowell, F.A.: Generalized entropy and the measurement of distributional change. Eur. Econ. Rev. **13**, 147–159 (1980a)

Cowell, F.A.: On the structure of additive inequality measures. Rev. Econ. Stud. **47**, 521–531 (1980b)

Cowell, F.A.: Measurement of inequality. In: Atkinson, A.B., Bourguignon, F. (eds.) Handbook of Income Distribution, vol. 1, pp. 87–166. North-Holland, Amsterdam (2000)

Cowell, F.A.: Measuring Inequality. Oxford University Press, New York (2011)

Cowell, F.A., Flachaire, E.: Statistical methods for distributional analysis. In: Atkinson, A.B., Bourguignon, F. (eds.) Handbook of Income Distribution, vol. 2A, pp. 359–465. North-Holland, Amsterdam (2015)

Cowell, F.A., Kuga, K.: Additivity and the entropy concept: an axiomatic approach to inequality measurement. J. Econ. Theory **25**, 131–143 (1981a)

Cowell, F.A., Kuga, K.: Inequality measurement: an axiomatic approach. Eur. Econ. Rev. **15**, 287–305 (1981b)

Cowell, F.A., Van Kerm, P.: Wealth inequality: a survey. J. Econ. Surv. **29**, 671–710 (2015)

Cowell, F.A., Jenkins, S.P., Litchfield, J.A.: The changing shape of the UK income distribution: kernel density estimates. In: Hills, J. (ed.) New Equalities: The Changing Distribution of Income and Wealth in the United Kingdom, pp. 49–75. Cambridge University Press, Cambridge (1996)

Cowell, F.A., Karagiannaki, E., McKnight, A.: Mapping and Measuring the Distribution of Household Wealth: A Cross-Country Analysis. LWS Working Paper 12, Luxembourg Income Study, Luxembourg (2012a). http://www.lisdatacenter.org/wps/lwswps/12.pdf

Cowell, F.A., Karagiannaki, E., McKnight, A.: Accounting for Cross-Country Differences in Wealth Inequality. LWS Working Paper 13, Luxembourg Income Study, Luxembourg (2012b). http://www.lisdatacenter.org/wps/lwswps/13.pdf

Cowell, F.A., Flachaire, E., Bandyopadhyay, S.: Reference distributions and inequality measurement. J. Econ. Inequal. **11**, 421–437 (2013)

Cowell, F.A., Davidson, R., Flachaire, E.: Goodness of fit: an axiomatic approach. J. Bus. Econ. Stat. **33**, 54–67 (2015)

Dabla-Norris, E., Kochhar, K., Suphaphiphat, N., Ricka, F., Tsounta, E.: Causes and Consequences of Income Inequality: A Global Perspective. Staff Discussion Note 15/13, International Monetary Fund, Washington, DC (2015). https://www.imf.org/external/pubs/cat/longres.aspx?sk=42986.0

D'Agostino, R.B., Stephens, M.A.: Goodness-of-Fit Techniques. Marcel Dekker, New York (1986)

Dagum, C.: A new model of personal income distribution: specification and estimation. Economie Appliquée **30**, 413–436 (1977)

Dagum, C.: The Distribution of Income and Wealth in Canada, the U.S.A. and Great Britain. In: Paper presented at the 4th Annual Convention of the Eastern Economic Association, Washington, DC (1978)
Dagum, C.: A model of net wealth distribution specified for negative, null and positive wealth. A case study: Italy. In: Dagum, C., Zenga, M. (eds.) Income and Wealth Distribution. Inequality and Poverty, pp. 42–56. Springer, Berlin (1990)
Dagum, C.: Human capital, income and wealth distribution models with applications. In: Proceedings of the Business and Economic Statistics Section, American Statistical Association, 154th Meeting, pp. 253–258 (1994)
Dagum, C.: Income and wealth distributions, Dagum system of. In: Balakrishnan, N., Read, C.B., Vidakovic, B. (eds.) Encyclopedia of Statistical Sciences, vol. 5, 2nd edn, pp. 3363–3378. Wiley, New York (2006a)
Dagum, C.: Wealth distribution models: analysis and applications. Statistica **66**, 235–268 (2006b)
Dalton, H.: The measurement of the inequality of incomes. Econ. J. **30**, 348–361 (1920)
Davies, J.B., Shorrocks, A.F.: The distribution of wealth. In: Atkinson, A.B., Bourguignon, F. (eds.) Handbook of Income Distribution, vol. 1, pp. 605–675. North-Holland, Amsterdam (2000)
Davies, J.B., Sandström, S., Shorrocks, A.F., Wolff, E.N.: The global pattern of household wealth. J. Int. Dev. **21**, 1111–1124 (2009)
Davies, J.B., Sandström, S., Shorrocks, A.F., Wolff, E.N.: The level and distribution of global household wealth. Econ. J. **121**, 223–254 (2011)
Deutsch, J., Fusco, A., Silber, J.: The BIP trilogy (bipolarization, inequality and polarization): one saga but three different stories. Econ.: Open-Access Open-Assess. E-J. 7, 1–33 (2013)
Drăgulescu, A., Yakovenko, V.M.: Exponential and power-law probability distributions of wealth and income in the United Kingdom and the United States. Phys. A: Stat. Mech. Appl. **299**, 213–221 (2001)
Espinguet, P., Terraza, M.: Essai d'extrapolation des distributions de salaires français. Economie Appliquée **36**, 535–561 (1983)
Esteban, J.M.: Income-share elasticity and the size distribution of income. Int. Econ. Rev. **27**, 439–444 (1986)
Esteban, J.-M., Ray, D.: On the measurement of polarization. Econometrica **62**, 819–851 (1994)
Esteban, M.D., Morales, D.: A summary on entropy statistics. Kybernetica **31**, 337–346 (1995)
Feng, S., Burkhauser, R.V., Butler, J.S.: Levels and long-term trends in earnings inequality: overcoming current population survey censoring problems using the GB2 distribution. J. Bus. Econ. Stat. **24**, 57–62 (2006)
Fitoussi, J.P.: Jean-Paul Fitoussi per lakrisi (2013). http://lakrisi.com/2013/01/11/jean-paul-fitoussi-per-lakrisi/
Fitoussi, J.P., Stiglitz, J.E.: The Ways Out of the Crisis and the Building of a More Cohesive World. Document de Travail de l'OFCE 17, Observatoire Français des Conjonctures Économiques, Paris (2009). http://www.ofce.sciences-po.fr/pdf/dtravail/WP2009-17.pdf
Foley, D.K.: A statistical equilibrium theory of markets. J. Econ. Theory **62**, 321–345 (1994)
Gallegati, M., Landini, S., Stiglitz, J.E.: The Inequality Multiplier, Mimeo (2015)
García, C., Callealta, F.J., Núñez, J.J.: La evolución de la distribución personal de la renta en España (1973–2000) a través de los parámetros del modelo de Dagum. El Trimestre Económico **292**, 783–808 (2006)
Gastwirth, J.L.: A general definition of the Lorenz curve. Econometrica **39**, 1037–1039 (1971)
Ghosh, J.K.: Higher Order Asymptotics. Institute of Mathematical Statistics and American Statistical Association, Hayward (1994)
Gibrat, R.: Les inégalités économiques. Applications: aux inégalités des richesses, à la concentration des entreprises, aux population des villes, aux statistiques des familles, etc., d'une loi nouvelle: la loi de l'effet proportionnel. Librairie du Recueil Sirey, Paris (1931)
Gini, C.: Sulla misura della concentrazione e della variabilità dei caratteri. Atti del Reale Istituto veneto di scienze, lettere ed arti 73, 1201–1248 (1914). English translation: On the measurement of concentration and variability of characters. Metron 63, 3–38 (2005)

Gottschalk, P., Smeeding, T.M.: Empirical evidence on income inequality in industrialized countries. In: Atkinson, A.B., Bourguignon, F. (eds.) Handbook of Income Distribution, vol. 1, pp. 261–307. Elsevier, New York (2000)

Gould, W.W., Pitblado, J., Poi, B.P.: Maximum Likelihood Estimation with Stata, 4th edn. Stata Press, College Station (2010)

Graf, M., Nedyalkova, D.: Modeling of income and indicators of poverty and social exclusion using the generalized beta distribution of the second kind. Rev. Income Wealth **60**, 821–842 (2014)

Hagerbaumer, J.B.: The Gini concentration ratio and the minor concentration ratio: a two parameter index of inequality. Rev. Econ. Stat. **59**, 377–379 (1977)

Hajargasht, G., Griffiths, W.E., Brice, J., Prasada, D.S., Rao, D.P., Chotikapanich, D.: Inference for income distributions using grouped data. J. Bus. Econ. Stat. **30**, 563–575 (2012)

Hardy, G.H., Littlewood, J.E., Pólya, G.: Some simple inequalities satisfied by convex functions. Mess. Math. **58**, 145–152 (1929)

Hegyi, G., Néda, Z., Santos, M.A.: Wealth distribution and Pareto's law in the Hungarian medieval society. Phys. A: Stat. Mech. Appl. **380**, 271–277 (2007)

Hickel, J.: A Short History of Neoliberalism (And How We Can Fix It). New Left Project (2012). http://www.newleftproject.org/index.php/site/article_comments/a_short_history_of_neoliberalism_and_how_we_can_fix_it

Jäntti, M., Jenkins, S.P.: The impact of macroeconomic conditions on income inequality. J. Econ. Inequal. **8**, 221–240 (2010)

Jäntti, M., Sierminska, E., Smeeding, T.M.: The Joint Distribution of Household Income and Wealth: Evidence from the Luxembourg Wealth Study. OECD Social, Employment and Migration Working Paper 65, Organisation for Economic Co-operation and Development, Paris (2008). http://www.oecd-ilibrary.org/social-issues-migration-health/the-joint-distribution-of-household-income-and-wealth_241506164527

Jäntti, M., Sierminska, E., Van Kerm, P.: The joint distribution of income and wealth. In: Gornick, J.C., Jäntti, M. (eds.) Income Inequality: Economic Disparities and the Middle Class in Affluent Countries, pp. 312–333. Stanford University Press, Stanford (2013)

Jäntti, M., Sierminska, E., Van Kerm, P.: Modelling the Joint Distribution of Income and Wealth. IZA Discussion Paper 9190, Institute for the Study of Labor, Bonn (2015). http://ftp.iza.org/dp9190.pdf

Jaynes, E.T.: Information theory and statistical mechanics. Phys. Rev. **106**, 620–630 (1957a)

Jaynes, E.T.: Information theory and statistical mechanics. II. Phys. Rev. **108**, 171–190 (1957b)

Jaynes, E.T.: Where do we stand on maximum entropy? In: Levine, R.D., Tribus, M. (eds.) The Maximum Entropy Formalism, pp. 15–118. MIT Press, Cambridge (1978)

Jenkins, S.P.: Distributionally-sensitive inequality indices and the GB2 income distribution. Rev. Income Wealth **55**, 392–398 (2009)

Jenkins, S.P., Jäntti, M.: Methods for Summarizing and Comparing Wealth Distributions. ISER Working Paper 2005–05, University of Essex, Colchester, UK (2005). https://www.iser.essex.ac.uk/publications/working-papers/iser/2005-05.pdf

Kakwani, N.: Income Inequality and Poverty: Methods of Estimation and Policy Applications. Oxford University Press, New York (1980)

Kaldor, N.: The irrelevance of equilibrium economics. Econ. J. **82**, 1237–1255 (1972)

Kaniadakis, G.: Non-linear kinetics underlying generalized statistics. Phys. A: Stat. Mech. Appl. **296**, 405–425 (2001)

Kaniadakis, G.: Statistical mechanics in the context of special relativity. Phys. Rev. E **66**, 056125 (2002)

Kaniadakis, G.: Statistical mechanics in the context of special relativity. II. Phys. Rev. E **72**, 036108 (2005)

Kaniadakis, G.: Maximum entropy principle and power-law tailed distributions. Eur. Phys. J. B **70**, 3–13 (2009)

Kaniadakis, G., Lissia, M., Scarfone, A.M.: Deformed logarithms and entropies. Phys. A: Stat. Mech. Appl. **340**, 41–49 (2004)

Kaniadakis, G., Lissia, M., Scarfone, A.M.: Two-parameter deformations of logarithm, exponential, and entropy: a consistent framework for generalized statistical mechanics. Phys. Rev. E **71**, 046128 (2005)

Kapur, J.N.: Maximum-Entropy Models in Science and Engineering. Wiley, New York (1989)

Khinchin, A.I.: Mathematical Foundations of Statistical Mechanics. Dover Publications, New York (1948)

Kirman, A.P.: Whom or what does the representative individual represent? J. Econ. Persp. **6**, 117–136 (1992)

Kirman, A.P., Koch, K.J.: Market excess demand in exchange economies with identical preferences and collinear endowments. Rev. Econ. Stud. **53**, 457–463 (1986)

Kleiber, C.: Dagum vs. Singh-Maddala income distributions. Econ. Lett. **53**, 265–268 (1996)

Kleiber, C.: The existence of population inequality measures. Econ. Lett. **57**, 39–44 (1997)

Kleiber, C.: A guide to the Dagum distributions. In: Chotikapanich, D. (ed.) Modeling Income Distributions and Lorenz Curves, pp. 97–117. Springer, New York (2008a)

Kleiber, C.: The Lorenz curve in economics and econometrics. In: Betti, G., Lemmi, A. (eds.) Advances on Income Inequality and Concentration Measures, pp. 225–242. Routledge, London (2008b)

Kleiber, C., Kotz, S.: Statistical Size Distributions in Economics and Actuarial Sciences. Wiley, New York (2003)

Kloek, T., van Dijk, H.K.: Efficient estimation of income distribution parameters. J. Econ. **8**, 61–74 (1978)

Kolm, S.-C.: Unequal inequalities. II. J. Econ. Theory **13**, 82–111 (1969)

Krämer, W., Ziebach, T.: The weak Pareto law and regular variation in the tails. Statistica **64**, 505–510 (2004)

Landini, S., Gallegati, M.: Heterogeneity, interaction and emergence: effects of composition. Int. J. Comput. Econ. Econom. **4**, 339–361 (2014)

Lebergott, S.: The shape of the income distribution. Am. Econ. Rev. **49**, 328–347 (1959)

Leipnik, R.B.: A maximum relative entropy principle for distribution of personal income with derivations of several known income distributions. Commun. Stat. Theory Methods **19**, 1003–1036 (1990)

Lorenz, M.O.: Methods of measuring the concentration of wealth. Publ. Am. Stat. Assoc. **9**, 209–219 (1905)

Luxembourg Income Study Database (LIS). http://www.lisdatacenter.org/ (multiple countries; July 1st-October 15th, 2015). LIS, Luxembourg

Luxembourg Wealth Study Database (LWS). http://www.lisdatacenter.org/ (multiple countries; August 6th-August 10th, 2015). LIS, Luxembourg

Lydall, H.F.: The Structure of Earnings. Clarendon Press, Oxford (1968)

Mandelbrot, B.: The Pareto-Lévy law and the distribution of income. Int. Econ. Rev. **1**, 79–106 (1960)

Mandelbrot, B.: Stable Paretian random functions and the multiplicative variation of income. Econometrica **29**, 517–543 (1961)

Marshall, A.: Principles of Economics, 8th edn. Macmillan, London (1920). http://oll.libertyfund.org/titles/1676

Marshall, A.W., Olkin, I.: Inequalities: Theory of Majorization and its Applications. Academic Press, New York (1979)

McDonald, J.B.: Some generalized functions for the size distribution of income. Econometrica **52**, 647–665 (1984)

McDonald, J.B., Ransom, M.R.: Functional forms, estimation techniques and the distribution of income. Econometrica **47**, 1513–1525 (1979)

McDonald, J.B., Ransom, M.R.: The generalized beta distribution as a model for the distribution of income: estimation of related measures of inequality. In: Chotikapanich, D. (ed.) Modeling Income Distributions and Lorenz Curves, pp. 147–166. Springer, New York (2008)

McDonald, J.B., Xu, Y.J.: A generalization of the beta distribution with applications. J. Econ. **66**, 133–152 (1995a)

McDonald, J.B.: YJ, Xu: Errata. J. Econ. **69**, 427–428 (1995b)

Milanovic, B.: True world income distribution, 1988 and 1993: first calculations based on household surveys alone. Econ. J. **112**, 51–92 (2002)

Monahan, J.F.: Numerical Methods of Statistics, 2nd edn. Cambridge University Press, New York (2011)

Murtin, F., Mira d'Ercole, M.: Household Wealth Inequality across OECD Countries: New OECD Evidence. OECD Statistics Brief 21, Organisation for Economic Co-operation and Development, Paris (2015). http://www.oecd.org/std/household-wealth-inequality-across-OECD-countries-OECDSB21.pdf

OECD. Growing Unequal? Income Distribution and Poverty in OECD Countries. OECD Publishing, Paris (2008)

OECD. Divided We Stand: Why Inequality Keeps Rising. OECD Publishing, Paris (2011)

OECD. In It Together: Why Less Inequality Benefits All. OECD Publishing, Paris (2015)

Okamoto, M.: Evaluation of the goodness of fit of new statistical size distributions with consideration of accurate income inequality estimation. Econ. Bull. **32**, 2969–2982 (2012a)

Okamoto, M.: Erratum to "Evaluation of the goodness of fit of new statistical size distributions with consideration of accurate income inequality estimation". Econ. Bull. **33**, 2443–2444 (2012b)

Okamoto, M.: Extension of the κ-Generalized Distribution: New Four-Parameter Models for the Size Distribution of Income and Consumption. LIS Working Paper 600, Luxembourg Income Study, Luxembourg (2013). http://www.lisdatacenter.org/wps/liswps/600.pdf

Ord, J.K.: Statistical models for personal income distribution. In: Patil, G.P., Kotz, S., Ord, J.K. (eds.) Statistical Distributions in Scientific Work, vol. 2, pp. 151–158. Reidel, Dordrecht (1975)

Ord, J.K., Patil, G.P., Taillie, C.: The choice of a distribution to describe personal incomes. In: Taillie, C., Patil, G.P., Baldessari, B.A. (eds.) Statistical Distributions in Scientific Work, vol. 6, pp. 193–201. D. Reidel Publishing Company, Dordrecht (1981)

Ostry, J.D., Berg, A.: Inequality and Unsustainable Growth: Two Sides of the Same Coin? Staff Discussion Note 11/8, International Monetary Fund, Washington, DC (2011). https://www.imf.org/external/pubs/cat/longres.aspx?sk=24686

Ostry, J.D., Berg, A., Tsangarides, C.G.: Redistribution, Inequality, and Growth. Staff Discussion Note 14/2, International Monetary Fund, Washington, DC (2014). https://www.imf.org/external/pubs/cat/longres.aspx?sk=41291.0

Papalexiou, S.M., Koutsoyiannis, D.: Entropy based derivation of probability distributions: a case study to daily rainfall. Adv. Water Resour. **45**, 51–57 (2012)

Pareto, V.: La legge della domanda. Giornale degli Economisti **10**, 59–68 (1895)

Pareto, V.: La courbe de la répartition de la richesse (1896). Reprinted in: Busino, G. (ed.) Œeuvres complètes de Vilfredo Pareto, Tome 3: Écrits sur la courbe de la répartition de la richesse, pp. 1–15. Librairie Droz, Geneva (1965)

Pareto, V.: Cours d'économie politique. Macmillan, London (1897a)

Pareto, V.: Aggiunta allo studio della curva delle entrate. Giornale degli Economisti **14**, 15–26 (1897b)

Park, S.Y., Bera, A.K.: Information Theoretic Approaches to Density Estimation with an Application to the U.S. Personal Income Data. Working Paper, Info-Metrics Institute, American University, Washington, DC (2010). http://www.american.edu/cas/economics/info-metrics/pdf/upload/Working-Paper-Park.pdf

Parzen, E.: Nonparametric statistical data modelling. J. Am. Stat. Assoc. **74**, 105–121 (1979)

Pérez, C.G., Alaiz, M.P.: Using the Dagum model to explain changes in personal income distribution. Appl. Econ. **43**, 4377–4386 (2011)

Pigou, A.C.: Wealth and Welfare. Macmillan, London (1912)

Piketty, T.: Book reviews. Economica **67**, 461–462 (2000)

Piketty, T.: Capital in the Twenty-First Century. The Belknap Press of Harvard University Press, Cambridge (2014)

Piketty, T., Zucman, G.: Wealth and inheritance in the long run. In: Atkinson, A.B., Bourguignon, F. (eds.) Handbook of Income Distribution, vol. 2B, pp. 1303–1368. North-Holland, Amsterdam (2015)
Pyatt, G., Chen, C.-N., Fei, J.: The distribution of income by factor components. Q. J. Econ. **95**, 451–473 (1980)
Quintano, C., D'Agostino, A.: Studying inequality in income distribution of sinlge-person households in four developed countries. Rev. Income Wealth **52**, 525–546 (2006)
Rajaonarison, D.: Deterministic heterogeneity in tastes and product differentiation in the K-logit model. Econ. Lett. **100**, 396–398 (2008)
Rajaonarison, D., Bolduc, D., Jayet, H.: The K-deformed multinomial logit model. Econ. Lett. **86**, 13–20 (2005)
Rao, C.: Linear Statistical Inference and its Applications. Wiley, New York (1973)
Ravallion, M.: The Luxembourg Income Study. J. Econ. Inequal. (2015). doi:10.1007/s10888-015-9298-y
Reed, W.J.: The Pareto law of incomes–an explanation and an extension. Phys. A: Stat. Mech. Appl. **319**, 469–486 (2003)
Reed, W.J.: On Pareto's law and the determinants of Pareto exponents. J. Income Distrib. **13**, 7–17 (2004)
Reed, W.J.: Brownian-Laplace motion and its use in financial modelling. Commun. Stat.: Theory. Methods **36**, 473–484 (2007)
Reed, W.J., Jorgensen, M.: The double Pareto-lognormal distribution–a new parametric model for size distributions. Commun. Stat.: Theory. Methods **33**, 1733–1753 (2004)
Reed, W.J., Wu, F.: New four- and five-parameter models for income distributions. In: Chotikapanich, D. (ed.) Modeling Income Distributions and Lorenz Curves, pp. 211–223. Springer, New York (2008)
Rodríguez, J.G., Salas, R.: Extended bi-polarization and inequality measures. Res. Econ. Inequal. **9**, 69–83 (2003)
Roine, J., Waldenström, D.: Long-run trends in the distribution of income and wealth. In: Atkinson, A.B., Bourguignon, F. (eds.) Handbook of Income Distribution, vol. 2A, pp. 469–592. North-Holland, Amsterdam (2015)
Rothschild, M., Stiglitz, J.E.: Increasing risk: I. A definition. J. Econ. Theory **2**, 225–243 (1970)
Sahota, G.S.: Theories of personal income distribution: a survey. J. Econ. Lit. **16**, 1–55 (1978)
Salem, A.B.Z., Mount, T.D.: A convenient descriptive model of income distribution: the gamma density. Econometrica **42**, 1115–1127 (1974)
Sargan, J.T.: The distribution of wealth. Econometrica **24**, 568–590 (1957)
Schneider, M.P.A.: A Comparative Entropy Analysis of the Distribution of Wages and Salaries, Mimeo (2008). http://www.peri.umass.edu/fileadmin/pdf/UM-NS_Workshop/NewSchool2008/MSchneider_-_Comparative_Entropy_Study_-08.pdf
Schneider, M.P.A.: Evidence for multiple labor market segments: an entropic analysis of US earned income, 1996–2007. J. Income Distrib. (2015)
Schoenberg, R.J.: Constrained maximum likelihood. Comput. Econ. **10**, 251–266 (1997)
Scott, D.W.: Multivariate Density Estimation: Theory, Practice, and Visualization, 2nd edn. Wiley, Hoboken (2015)
Shannon, C.E.: A mathematical theory of communication. Bell Syst. Tech. J. **27**, 379–423 and 623–657 (1948)
Shorrocks, A.F.: The class of additively decomposable inequality measures. Econometrica **48**, 613–625 (1980)
Shorrocks, A.F.: Ranking income distributions. Economica **50**, 3–17 (1983)
Sierminska, E., Smeeding, T.: Measurement Issues: Equivalence Scales, Accounting Framework, and Reference Unit. In: Paper presented at the LWS Workshop: "Construction and Usage of Comparable Microdata on Wealth: The LWS", Perugia, Italy (2005). http://www.researchgate.net/profile/Eva_Sierminska/publication/237340989_Measurement_Issues_Equivalence_Scales_Accounting_Framework_and_Reference_Unit/links/00b7d531585c76e93e000000.pdf

Sierminska, E., Brandolini, A., Smeeding, T.M.: Comparing Wealth Distribution across Rich Countries: First Results from the Luxembourg Wealth Study. LWS Working Paper 1, Luxembourg Income Study, Luxembourg (2006a). http://www.lisdatacenter.org/wps/lwswps/1.pdf
Sierminska, E., Brandolini, A., Smeeding, T.M.: The Luxembourg Wealth Study: a cross-country comparable database for household wealth research. J. Econ. Inequal. **4**, 375–383 (2006b)
Sierminska, E., Smeeding, T.M., Allegrezza, S.: The distribution of assets and debt. In: Gornick, J.C., Jäntti, M. (eds.) Income Inequality: Economic Disparities and the Middle Class in Affluent Countries, pp. 285–311. Stanford University Press, Stanford (2013)
Singh, S.K., Maddala, G.S.: A function for size distribution of incomes. Econometrica **44**, 963–970 (1976)
Stanley, M.H.R., Buldyrev, S.V., Havlin, S., Mantegna, R.N., Salinger, M.A., Stanley, H.E.: Zipf plots and the size distribution of firms. Econ. Lett. **49**, 453–457 (1995)
Stich, A.: Inequality and negative income. J. Ital. Stat. Soc. **5**, 297–305 (1996)
Stiglitz, J.E.: Distribution of income and wealth among individuals. Econometrica **37**, 382–397 (1969)
Stiglitz, J.E.: The Price of Inequality: How Today's Divided Society Endangers Our Future. W. W. Norton & Company, New York (2012)
Stiglitz, J.E.: The Great Divide: Unequal Societies and What We Can Do About Them. W. W. Norton & Company, New York (2015a)
Stiglitz, J.E.: New Theoretical Perspectives on the Distribution of Income and Wealth among Individuals: Part I: The Wealth Residual. Working Paper 21189, National Bureau of Economic Research, Cambridge, MA, (2015b). http://www.nber.org/papers/w21189
Stiglitz, J.E.: New Theoretical Perspectives on the Distribution of Income and Wealth among Individuals: Part II: Equilibrium Wealth Distributions. Working Paper 21190, National Bureau of Economic Research, Cambridge, MA (2015c). http://www.nber.org/papers/w21190
Stiglitz, J.E.: New Theoretical Perspectives on the Distribution of Income and Wealth among Individuals: Part III: Life Cycle Savings vs. Inherited Savings. Working Paper 21191, National Bureau of Economic Research, Cambridge, MA (2015d). http://www.nber.org/papers/w21191
Stiglitz, J.E.: New Theoretical Perspectives on the Distribution of Income and Wealth among Individuals: Part IV: Land and Credit. Working Paper 21192, National Bureau of Economic Research, Cambridge, MA (2015e). http://www.nber.org/papers/w21192
Stiglitz, J.E., Ocampo, J.A., Spiegel, S., Ffrench-Davis, R., Nayyar, D.: Stability with Growth: Macroeconomics. Liberalization and Development. Oxford University Press, New York (2006)
Sturges, H.A.: The choice of a class interval. J. Am. Stat. Assoc. **21**, 65–66 (1926)
Summers, L.H.: The scientific illusion in empirical macroeconomics. Scand. J. Econ. **3**, 129–148 (1991)
Tachibanaki, T., Suruga, T., Atoda, N.: Estimations of income distribution parameters for individual observations by maximum likelihood method. J. Jpn. Stat. Soc. **27**, 191–203 (1997)
Taillie, C.: Lorenz ordering within the generalized gamma family of income distributions. In: Taillie, C., Patil, G.P., Baldessari, B.A. (eds.) Statistical Distributions in Scientific Work, vol. 6, pp. 181–192. D. Reidel Publishing Company, Dordrecht (1981)
Takayasu, H.: Fractals in the Physical Sciences. Manchester University Press, Manchester (1990)
Theil, H.: Economics and Information Theory. North-Holland, Amsterdam (1967)
Therborn, G.: The Killing Fields of Inequality. Polity Press, Cambridge (2013)
Thode, H.C.: Testing for Normality. Marcel Dekker, New York (2002)
Van Kerm, P.: Adaptive kernel density estimation. Stata J. **3**, 148–156 (2003)
Vaughan, R.N.: A Study of the Distribution of Wealth. Ph.D. Thesis, University of Cambridge, Cambridge (1975)
Venter, G.: Transformed beta and gamma distributions and aggregate losses. Proc. Casualty Actuar. Soc. **70**, 156–193 (1983)
Wilfling, B.: Lorenz ordering of generalized beta-II income distributions. J. Econ. **71**, 381–388 (1996a)
Wilfling, B.: A sufficient condition for Lorenz ordering. Sankhyā: Indian. J Stat. **58**, 62–69 (1996b)

Wilfling, B., Krämer, W.: The Lorenz ordering of Singh-Maddala income distributions. Econ. Lett. **43**, 53–57 (1993)

Wold, H.O.A., Whittle, P.: A model explaining the Pareto distribution of wealth. Econometrica **25**, 591–595 (1957)

Wolfson, M.C.: When inequalities diverge. Am. Econ. Rev. **84**, 353–358 (1994)

Wu, X.: Calculation of maximum entropy densities with application to income distribution. J. Econ. **115**, 347–354 (2003)

Wu, X., Perloff, J.M.: China's income distribution, 1985–2001. Rev. Econ. Stat. **87**, 763–775 (2005)

Yule, G.U.: A mathematical theory of evolution, based on the conclusions of Dr. J. C. Willis, F.R.S. Philos. Trans. Royal Soc. B: Biol. Sci. **213**, 21–87 (1925)

Index of Citations

F. Clementi and M. Gallegati, *The Distribution of Income and Wealth*,
New Economic Windows, DOI 10.1007/978-3-319-27410-2

MIX
Papier aus verantwortungsvollen Quellen
Paper from responsible sources
FSC® C105338

If you have any concerns about our products,
you can contact us on
ProductSafety@springernature.com

In case Publisher is established outside the EU,
the EU authorized representative is:
Springer Nature Customer Service Center GmbH
Europaplatz 3, 69115 Heidelberg, Germany

Printed by Libri Plureos GmbH
in Hamburg, Germany